Lecture Notes in Control and Information Sciences

Edited by M. Thoma and A. Wyner

Vol. 81: Stochastic Optimization
Proceedings of the International Conference, Kiew, 1984
Edited by I. Arkin, A. Shiraev, R. Wets
X, 754 pages, 1986.

Vol. 82: Analysis and Algorithms of Optimization Problems
Edited by K. Malanowski, K. Mizukami
VIII, 240 pages, 1986.

Vol. 83: Analysis and Optimization of Systems
Proceedings of the Seventh International Conference of Analysis and Optimization of Systems
Antiba, June 26-27, 1986
Edited by A. Bensoussan, J. L. Lions
XVI, 901 pages, 1986.

Vol. 84: System Modelling and Optimization
Proceedings of the 12th IFIP Conference
Budapest, Hungary, September 2–6, 1985
Edited by A. Prékopa, J. Szelezsán, B. Strazicky
XII, 1046 pages, 1986.

Vol. 85: Stochastic Processes in Underwater Acoustics
Edited by Charles R. Baker
V, 205 pages, 1986.

Vol. 86: Time Series and Linear Systems
Edited by Sergio Bittanti
XVII, 243 pages, 1986.

Vol. 87: Recent Advances in System Modelling and Optimization
Proceedings of the IFIP-WG 7/1 Working Conference
Santiago, Chile, August 27-31, 1984
Edited by L. Contesse, R. Correa, A. Weintraub
IV, 199 pages, 1987.

Vol. 88: Bruce A. Francis
A Course in H_∞ Control Theory
XI, 156 pages, 1987.

Vol. 88: Bruce A. Francis
A Course in H_∞ Control Theory
X, 150 pages, 1987.
Corrected - 1st printing 1987

Vol. 89: G. K. H. Pang/A. G. J. McFarlane
An Expert System Approach to Computer-Aided Design of Multivariable Systems
XII, 223 pages, 1987.

Vol. 90: Singular Perturbations and Asymptotic Analysis in Control Systems
Edited by P. Kokotovic, A. Bensoussan, G. Blankenship
VI, 419 pages, 1987.

Vol. 91 Stochastic Modelling and Filtering
Proceedings of the IFIP-WG 7/1 Working Conference
Rome, Italy, Decembre 10-14, 1984
Edited by A. Germani
IV, 209 pages, 1987.

Vol. 92: L. T. Grujić, A. A. Martynyuk, M. Ribbens-Pavella
Large-Scale Systems Stability Under Structural and Singular Perturbations
XV, 366 pages, 1987.

Vol. 93: K. Malanowski
Stability of Solutions to Convex Problems of Optimization
IX, 137 pages, 1987.

Vol. 94: H. Krishna
Computational Complexity of Bilinear Forms
Algebraic Coding Theory and Applications to Digital Communication Systems
XVIII, 166 pages, 1987.

Vol. 95: Optimal Control
Proceedings of the Conference on Optimal Control and Variational Calculus
Oberwolfach, West-Germany, June 15-21, 1986
Edited by R. Bulirsch, A. Miele, J. Stoer and K. H. Well
XII, 321 pages, 1987.

Vol. 96: H. J. Engelbert/W. Schmidt
Stochastic Differential Systems
Proceedings of the IFIP-WG 7/1 Working Conference
Eisenach, GDR, April 6-13, 1986
XII, 381 pages, 1987.

For information about Vols. 1–80 please contact your bookseller or Springer-Verlag.

Lecture Notes in
Control and
Information Sciences

Edited by M. Thoma and A. Wyner

146

D. Mustafa, K. Glover

Minimum Entropy H_∞ Control

Springer-Verlag
Berlin Heidelberg GmbH

Authors
Denis Mustafa
Laboratory for Information
and Decision Systems
Massachusetts Institute for
Technology
Cambridge MA 02139, USA

Keith Glover
Department of Engineering
Cambridge University
Cambridge CB2 1PZ, UK

ISBN 978-3-540-52947-7 ISBN 978-3-540-47182-0 (eBook)
DOI 10.1007/978-3-540-47182-0

Originally published by Springer-Verlag Berlin Heidelberg New York in 1990.

2161/3020-543210 Printed on acid-free paper.

To my parents

D.M.

Preface

Optimal control of linear dynamic systems by linear controllers generally reduces to minimizing some norm of a closed-loop transfer function over the class of all stabilizing controllers. Different norms can be motivated by different assumptions on the physical system and signals. Robust control then attempts to maintain good performance—such as attenuation of disturbances—in the presence of uncertainty in the plant. One realistic way of modelling plant uncertainty is to include perturbations with bounded $\mathcal{H}_\infty$-norm. In this context it is natural to require controllers to make particular closed-loop transfer functions have $\mathcal{H}_\infty$-norm less than some prescribed tolerance. The central question addressed in this monograph is to investigate which of these admissible closed-loop transfer functions to choose. One possible choice is to take that which minimizes an *entropy integral* and the derivation and properties of this solution occupies most of this monograph. The approach taken here exploits recent progress in the state-space solution to the $\mathcal{H}_\infty$ control problem and explicit formulae can be obtained for the solution.

At a mathematical level the entropy minimization problem considered here has close connections with maximum entropy spectral analysis. However the objective is more easily interpreted as a compromise between an $\mathcal{H}_\infty$ constraint and a Linear Quadratic Gaussian objective, rather than in terms of information theory or of one of the many other uses of the word entropy. Relations to other control methods are also exposed here.

The appropriate background for this monograph is a first course in state-space methods, together with an elementary knowledge of $\mathcal{H}_\infty$ control and Linear Quadratic Gaussian control.

The bulk of the research reported in this monograph was carried out at the Engineering Department at Cambridge University, UK, between October 1986 and May 1989. Financial support was provided by the Science and Engineering Research Council. Since September 1989 the first author has been a Harkness Fellow at the Laboratory for Information and Decision Systems at the Massachusetts Institute of Technology, USA, where the final stages of this research were carried out. Financial support was provided by the Air Force Office of Scientific Research, grant number AFOSR-89-0276.

The first author would like to thank Dr Raimund Ober for his sound advice and for many useful discussions and comments; Glenn Vinnicombe for some influential discussions regarding Chapter 5; and Dr Duncan McFarlane, Pablo Iglesias, James Sefton, Frank van Diggelen, Richard Hyde and Paul McGinnie for their comments on earlier versions of the manuscript. Thanks too to Dr Basil Kouvaritakis, for sparking off the first author's interest in multivariable control.

DENIS MUSTAFA, *Cambridge, Massachusetts, USA*

KEITH GLOVER, *Cambridge, UK*

Contents

List of Figures

Chapter 1

Introduction

1.1 Overview

In control system analysis and design the first step is to produce a mathematical model of the system being controlled. This model includes not only the dynamic equations of the system but also a characterization of any disturbances affecting the system. If the dynamic model is assumed to be exact and the disturbances are assumed to be Gaussian stochastic processes, then the control law that minimizes the expected value of a quadratic form in the errors is given by the Linear Quadratic Gaussian (LQG) controller [2, 38]. However there is no guaranteed *robustness* [15] for LQG-controllers, and a small perturbation of the plant dynamics may give rise to an unstable closed-loop system. The difficulty is that the design method concentrates on the gain of the closed-loop system to the disturbance signals and does not explicitly consider uncertainty in the dynamic model.

A reliable control synthesis procedure should therefore take account of the *uncertainty* inherent in a mathematical model [18]. A suitable norm that can incorporate both signal gain and robustness to plant uncertainty is the $\mathcal{H}_\infty$-norm. The synthesis of controllers using $\mathcal{H}_\infty$-optimal control methods was initiated in [67] and a wide range of sensible robust controller synthesis problems may be phrased in the $\mathcal{H}_\infty$-optimal control framework (see for example [24, 40]). If the uncertainty has a particular structure or robust performance is required then structured singular values need to be considered as in [19]. However a central constituent of all these problems is to characterize controllers that satisfy a bound on the $\mathcal{H}_\infty$-norm of a (usually frequency weighted) closed-loop transfer function. Advances in the algorithms [28, 17, 30] have now reduced the computational burden of these methods to one approaching that of the LQG-optimal control problem.

The family of controllers that satisfy a closed-loop $\mathcal{H}_\infty$-norm bound is characterized by a linear fractional transformation of a 'ball in $\mathcal{H}_\infty$' and a natural question is which element of this ball to choose. One choice that has been considered in a closely related problem in mathematics is to choose that which minimizes an entropy integral [3, 4, 21, 22]. The full development of the minimum entropy solution to the present control problem, particularly exploiting state-space techniques, is the central problem considered in this monograph. This minimum entropy solution is a precisely defined compromise between the LQG-optimal solution and the $\mathcal{H}_\infty$-optimal solution, which are themselves obtained when the $\mathcal{H}_\infty$-norm constraint is relaxed or given its minimum value.

1.2 Outline of this Monograph

An outline of the contents of this monograph is now given.

Chapter 2: The Entropy of a System We introduce the *entropy* of a system satisfying an $\mathcal{H}_\infty$-norm bound, and give some properties. In particular, the *entropy at infinity* is shown to be an upper bound on the more familiar *LQG cost.* The $\mathcal{H}_\infty$-norm bound on the system is given its usual *robustness* interpretation. The entropy itself and the properties and interpretations described above are vital components of the remaining chapters.

Chapter 3: The Minimum Entropy $\mathcal{H}_\infty$ Control Problem In this chapter the *Minimum Entropy $\mathcal{H}_\infty$ Control Problem* is posed and solved. The problem is to determine a controller which minimizes the entropy of a system whilst stabilizing it and satisfying a closed-loop $\mathcal{H}_\infty$-norm bound. The problem is posed in the same general framework used in $\mathcal{H}_\infty$-optimal control, into which a wide range of useful control problems may be embedded. When entropy is evaluated at infinity, a particularly simple and explicit solution is obtained, and state-space formulae are derived. By exploiting the relationship between entropy and LQG cost, an inherent *$\mathcal{H}_\infty$/LQG tradeoff* is proved. Furthermore, if the $\mathcal{H}_\infty$ constraint is relaxed entirely in the Minimum Entropy $\mathcal{H}_\infty$ Control Problem, then the corresponding LQG control problem is recovered.

Chapter 4: The Minimum Entropy $\mathcal{H}_\infty$ Distance Problem The $\mathcal{H}_\infty$ *General Distance Problem* of finding all causal extensions of a certain anticausal transfer function within a prescribed distance is an interesting problem in its own right, as well as being (until recently) a stepping-stone to the solution of $\mathcal{H}_\infty$-optimal control problems. We derive the minimum entropy solution. This forms an alternative solution to the Minimum Entropy $\mathcal{H}_\infty$ Control Problem of the previous chapter.

Chapter 5: Relations to Combined $\mathcal{H}_\infty$/LQG Control We have by now seen how the Minimum Entropy $\mathcal{H}_\infty$ Control Problem combines both $\mathcal{H}_\infty$ and LQG criteria. Other researchers have proposed apparently quite different methods for combining $\mathcal{H}_\infty$ and LQG criteria. One such *Combined $\mathcal{H}_\infty$/LQG Problem* minimizes an *auxiliary LQG cost.* We prove that, under certain conditions, the auxiliary cost and the entropy are identical, so then the Combined $\mathcal{H}_\infty$/LQG Problem and the Minimum Entropy $\mathcal{H}_\infty$ Control Problem coincide. This result has the bonus of allowing a simplification of the solution to the Combined $\mathcal{H}_\infty$/LQG Problem, by using the state-space solution of the Minimum Entropy $\mathcal{H}_\infty$ Control Problem derived in Chapter 3.

Chapter 6: Relations to Risk-Sensitive LQG Control For completeness, the equivalence between the Minimum Entropy $\mathcal{H}_\infty$ Control Problem and the *Risk-Sensitive LQG Control Problem* is reviewed. Risk-sensitive LQG control is a generalization of standard LQG control to include an *exponential-of-quadratic cost.* We prove that, under certain conditions, the exponential-of-quadratic cost and the entropy are identical, so then the Risk-Sensitive LQG Control Problem and the Minimum Entropy $\mathcal{H}_\infty$ Control Problem coincide.

Chapter 7: The Normalized $\mathcal{H}_\infty$ Control Problem At the end of Chapter 3, it was shown that an LQG control problem may be recovered from each Minimum Entropy $\mathcal{H}_\infty$ Control Problem by relaxing the $\mathcal{H}_\infty$ constraint entirely. Here this reasoning is reversed: beginning with the well-known Normalized LQG Problem, we invoke the minimum entropy/$\mathcal{H}_\infty$ constraint to obtain the corresponding (minimum entropy) *Normalized $\mathcal{H}_\infty$ Control Problem*, and in the process gain robustness at the expense of the LQG cost bound. This Normalized $\mathcal{H}_\infty$ Control Problem is shown to be a sensible one, and a numerical example is given to illustrate the tradeoffs.

Chapter 8: $\mathcal{H}_\infty$-Characteristic Values Internally *balanced* state-space realizations are a well-known route to model reduction. We firstly review *LQG-balancing* of a system, where a given system is balanced in a closed-loop with a Normalized LQG Controller. Invoking the ideas of the previous chapter, we then provide a natural minimum entropy/$\mathcal{H}_\infty$ generalization: *$\mathcal{H}_\infty$-balancing*. Here the system is balanced in a closed-loop with a Normalized $\mathcal{H}_\infty$ Controller (from Chapter 7). This allows the definition of a new set of input-output invariants, called the *$\mathcal{H}_\infty$-characteristic values*. Hence a new model reduction scheme—*$\mathcal{H}_\infty$-balanced truncation*—is proposed and its properties are explored. Some of the main features are highlighted in a short numerical example. Strong connections with *coprime factorization* are derived along the way.

Chapter 9: LQG and $\mathcal{H}_\infty$ Monotonicity In Chapter 3 an $\mathcal{H}_\infty$/LQG tradeoff was proved: as the $\mathcal{H}_\infty$ constraint is relaxed (hence as robustness guarantees loosen) so the entropy decreases. Here we explore two stronger properties displayed by many minimum entropy $\mathcal{H}_\infty$ controllers, that of *LQG monotonicity* and *$\mathcal{H}_\infty$ monotonicity:* as the $\mathcal{H}_\infty$ constraint is relaxed, the LQG cost *decreases* monotonically, whereas the $\mathcal{H}_\infty$-norm *increases* monotonically. We conjecture that any minimum entropy/$\mathcal{H}_\infty$ controller is both LQG monotonic and $\mathcal{H}_\infty$ monotonic. An expression for the derivative of the LQG cost is derived, based on the state-space solution to the Minimum Entropy $\mathcal{H}_\infty$ Distance Problem from Chapter 4. This is a promising direction for the construction of a proof of LQG monotonicity in general.

Appendix A: Proof of Results Needed in the Text To preserve the flow of the main body of the text, some of the proofs are presented here.

Appendix B: Entropy Formulae: Alternative Derivation For added insight, an alternative derivation of the entropy formulae and some related results from Chapter 3 is presented.

Appendix C: Notation Although notation is standard as far as possible, for clarity the main notational conventions are listed.

1.3 Background References

The reader is assumed to have a grounding in the fundamentals of the following three areas.

- **Essential system theory,** such as controllability, observability and related concepts. A comprehensive reference is [36, Chapters 1 & 2].
- **LQG control theory,** as covered by [2] or [38], for example.
- **$\mathcal{H}_\infty$-optimal control theory,** as covered by [24]. We shall assume a basic familiarity with Hardy spaces, but not beyond that contained in [24, Chapter 2].

Each of these three areas is now sufficiently mature to have a vast literature. We shall therefore restrict review material to the appropriate parts of chapters and cite essential references only.

Chapter 2

The Entropy of a System

2.1 Introduction

Entropy has established itself as an important notion, with a wide applicability in a number of diverse subjects. For example, Shannon [54] introduced the concept of entropy as an information measure, whilst Burg [11] proposed a successful entropy method in spectral analysis. In this chapter we introduce the entropy of a system which satisfies an $\mathcal{H}_\infty$-norm bound, and derive some important properties. We show that the entropy is an upper bound on the LQG cost, and interpret the $\mathcal{H}_\infty$-norm bound on the system as providing a prespecified level of robustness.

2.2 Definition of the Entropy

Let us begin with a formal definition.

Definition 2.2.1 (Entropy at $s_0 \in (0,\infty)$) *Let $H \in \mathcal{RL}_\infty$ and let $\gamma \in \mathbb{R}$ be such that $\|H\|_\infty < \gamma$. Let $s_0 \in (0,\infty)$. Then the* entropy of H at s_0 *is defined by*

$$I(H;\gamma;s_0) := -\frac{\gamma^2}{2\pi}\int_{-\infty}^{\infty} \ln|\det(I-\gamma^{-2}H^*(j\omega)H(j\omega))|\left[\frac{s_0^2}{s_0^2+\omega^2}\right]d\omega.$$

Apart from an extra factor of s_0 in the numerator, this definition is equivalent to the entropy integral defined in [4, 3] in the context of indeterminate extension problems. This extra factor allows us to extend the domain of definition to include the case when $s_0 \to \infty$. In doing so, we obtain the *entropy at infinity* of the system, which is of great significance in what is to follow.

Definition 2.2.2 (Entropy at infinity) *Let $H \in \mathcal{RL}_\infty$ and let $\gamma \in \mathbb{R}$ be such that $\|H\|_\infty < \gamma$. Then the* entropy of H at infinity *is defined by*

$$I(H;\gamma;\infty) := \lim_{s_0\to\infty}\{I(H;\gamma;s_0)\}.$$

Remark 2.2.3 The entropy of H is a useful measure of how close H is to the upper bound γ on $\sigma_1\{H(j\omega)\}$. This can be most clearly seen by rewriting the entropy as

$$I(H;\gamma;s_0) = -\frac{\gamma^2}{2\pi}\int_{-\infty}^{\infty}\sum_i \ln|1-\gamma^{-2}\sigma_i^2\{H(j\omega)\}|\left[\frac{s_0^2}{s_0^2+\omega^2}\right]d\omega. \quad (2.1)$$

If $\sigma_1^2\{H(j\omega)\} > \gamma^2-\epsilon^2$ for some frequency range $\omega_1 < \omega < \omega_2$, then $I(H;\gamma;s_0) \to \infty$ as $\epsilon \to 0$. Also, it is clear from (2.1) that all the singular values $\sigma_i\{H\}$ of H are included, unlike the $\mathcal{H}_\infty$-norm which depends only on the largest singular value $\sigma_1\{H\}$.

Remark 2.2.4 By convention, in future we will use 'entropy' to mean 'entropy at infinity' unless otherwise stated.

Remark 2.2.5 The domain of definition of the entropy at s_0 can in fact be extended to the whole of the right-half plane by replacing the term $s_0^2/(s_0^2+\omega^2)$ with $(\mathrm{Re}\{s_0\})^2/(|s_0-j\omega|)^2$, but we shall not need to do this. (To look ahead a little, the control systems derived in later chapters are physically realizable (real-rational) only if s_0 is real.)

As a glimpse of the importance of the entropy, we mention here some of its interpretations. We will see further on in this chapter precisely how to relate entropy to the LQG cost associated with a system. In Chapter 5, we outline the combined $\mathcal{H}_\infty$/LQG framework of [8, 9] in which an *auxiliary* LQG cost is used. We prove that, under certain conditions, the auxiliary cost is exactly the entropy. In Chapter 6 we cover the equivalence between entropy and the exponential-of-quadratic cost of the Risk-Sensitive LQG Control Problem of [34, 57]. The entropy will also be interpreted in the context of 'band extension' problems [20] in Chapter 4.

2.3 Some Properties

We now state various properties of the entropy, some of which will be needed later.

Proposition 2.3.1 *Let $H \in \mathcal{RL}_\infty$ and let $\gamma \in \mathbb{R}$ be such that $\|H\|_\infty < \gamma$. Let $s_0 \in (0, \infty]$. Then*

(i) *$I(H;\gamma;s_0)$ is well-defined.*

(ii) *$I(H;\gamma;s_0) \geq 0$.*

(iii) *$I(H;\gamma;s_0) < \infty$ for any $s_0 \in (0,\infty)$.*

(iv) *$I(H;\gamma;\infty) < \infty$ if and only if $H(\infty) = 0$.*

(v) *$I(H;\gamma;s_0) = 0$ if and only if $H = 0$.*

(vi) *$I(UHV;\gamma;s_0) = I(H;\gamma;s_0)$ for any U, $V \in \mathcal{RL}_\infty$ such that $U^*U = I$, $VV^* = I$.*

(vii) *$I(H^*;\gamma;s_0) = I(H^T;\gamma;s_0) = I(H;\gamma;s_0)$.*

Proof For convenience, define

$$\alpha(\gamma;\omega) := -\frac{\gamma^2}{2\pi}\ln|\det(I - \gamma^{-2}H^*(j\omega)H(j\omega))| \tag{2.2}$$

and

$$\beta(\omega;s_0) := \frac{s_0^2}{s_0^2+\omega^2} \tag{2.3}$$

so that

$$I(H;\gamma;s_0) = \int_{-\infty}^{\infty} \alpha(\gamma;\omega)\beta(\omega;s_0)\,d\omega. \tag{2.4}$$

Begin with the fact that

$$\|H\|_\infty < \gamma \quad \Longleftrightarrow \quad I \geq I - \gamma^{-2}H^*(j\omega)H(j\omega) > 0, \qquad \forall \omega \in \mathbb{R} \cup \{\infty\}. \tag{2.5}$$

This implies that

$$0 \leq \alpha(\gamma;\omega) < b < \infty, \qquad \forall \omega \in \mathbb{R} \cup \{\infty\} \text{ and for some } b \in \mathbb{R}. \tag{2.6}$$

Also, it is obvious that

$$0 \leq \beta(\omega; s_0) \leq 1, \qquad \forall \omega \in \mathbb{R} \cup \{\infty\} \text{ and } \forall s_0 \in (0,\infty].$$

Hence the integral is well-defined, which proves Part (i), and non-negative, which proves Part (ii). Now if $s_0 \in (0,\infty)$, we also have

$$(1+\omega^2)\beta(\omega; s_0) \to s_0^2 < \infty \quad \text{as} \quad \omega \to \pm\infty.$$

Part (iii) follows from this together with (2.6). For Part (iv) we note that, as $s_0 \to \infty$, we have $\beta(\omega; s_0) \to 1$ monotonically from below for each $\omega \in \mathbb{R}$. Now, because H is rational and proper, $\omega\sigma_1\{H(j\omega)\}$ is uniformly bounded if and only if $H(\infty) = 0$. It can then be shown, by using Lemma A.2.1(i) of Appendix A, that

$$\begin{aligned} H(\infty) = 0 \quad &\Longleftrightarrow \\ &(1+\omega^2)\alpha(\gamma;\omega) < c < \infty, \quad \forall \omega \in \mathbb{R} \cup \{\infty\} \text{ and for some } c \in \mathbb{R}. \end{aligned} \tag{2.7}$$

Clearly, when H satisfies (2.7), $\alpha(\gamma;\omega)$ is integrable and $I(H;\gamma;\infty)$ is finite by dominated convergence, which proves Part (iv). Part (v) is a simple consequence of the fact that the integrand $\alpha(\gamma;\omega)\beta(\omega;s_0)$ is non-negative. Parts (vi) and (vii) are immediate on using the well-known matrix identity

$$\det(I - M^*M) = \det(I - MM^*) = \det(I - (M^T)^*M^T).$$

□

The next proposition shows that the entropy at $s_0 \in (0,\infty]$ of a fixed $H \in \mathcal{RL}_\infty$ is monotonically decreasing with γ. This will be useful in Section 3.6.

Proposition 2.3.2 *Let $H \in \mathcal{RL}_\infty$ and let $\gamma \in \mathbb{R}$ be such that $\|H\|_\infty < \gamma$. If $s_0 \in (0,\infty)$ then $I(H;\gamma;s_0)$ is a monotonically decreasing function of γ. If $H(\infty) = 0$ then $I(H;\gamma;\infty)$ is a monotonically decreasing function of γ.*

Proof Firstly note that, by Proposition 2.3.1(iii) and (iv), $I(H;\gamma;s_0)$ and $I(H;\gamma;\infty)$ are both finite under the stated assumptions on H.

Define $\xi := \gamma^{-2}$ and let $\epsilon \in \mathbb{R}$. It is easy to show that

$$\begin{aligned} -\ln|\det(I - (\xi+\epsilon)H^*(j\omega)H(j\omega))| &+ \ln|\det(I - \xi H^*(j\omega)H(j\omega))| \\ &= -\ln|\det(I - \epsilon(I - \xi H^*(j\omega)H(j\omega))^{-1}H^*(j\omega)H(j\omega))| \\ &= \epsilon \operatorname{trace}[(I - \xi H^*(j\omega)H(j\omega))^{-1}H^*(j\omega)H(j\omega)] + O(\epsilon^2), \end{aligned}$$

using Lemma A.2.1(i) of Appendix A to get the second equality. It follows that

$$\begin{aligned}\frac{\partial}{\partial \xi}\{-\ln|\det(I-\xi H^*(j\omega)H(j\omega))|\}\\ = \operatorname{trace}[(I-\xi H^*(j\omega)H(j\omega))^{-1}H^*(j\omega)H(j\omega)].\end{aligned}$$

Therefore, continuing with the notation of equations (2.2), (2.3) and (2.4), we have

$$\begin{aligned}\frac{\partial}{\partial \gamma}\alpha(\gamma;\omega) &= \frac{\partial}{\partial \xi}\left\{-\frac{1}{2\pi\xi}\ln|\det(I-\xi H^*(j\omega)H(j\omega))|\right\}\frac{\partial \xi}{\partial \gamma}\\ &= -\frac{1}{\pi\xi\gamma}\ln|\det(I-\xi H^*(j\omega)H(j\omega))|\\ &\quad -\frac{1}{\pi\xi\gamma}\operatorname{trace}[(I-\xi H^*(j\omega)H(j\omega))^{-1}\xi H^*(j\omega)H(j\omega)]. \qquad (2.8)\end{aligned}$$

Define $\kappa_i := \lambda_i\{\xi H^*(j\omega)H(j\omega)\}$, and note that $0 \le \kappa_i < 1$ because $\|H\|_\infty < \gamma$. We can then rewrite (2.8) as

$$\frac{\partial}{\partial \gamma}\alpha(\gamma;\omega) = -\frac{\gamma}{\pi}\sum_i\left(\ln(1-\kappa_i)+(1-\kappa_i)^{-1}\kappa_i\right). \qquad (2.9)$$

Define $f(x) := \ln(1-x)+(1-x)^{-1}x$ for $0 \le x < 1$. It is straightforward to show that $df/dx = (1-x)^{-2}x \ge 0$ for $0 \le x < 1$, so $f(x)$ is a monotonically increasing function of x. But $f(0) = 0$ so $f(x) \ge 0$ for $0 \le x < 1$. Applying this to (2.9) gives

$$\frac{\partial}{\partial \gamma}\alpha(\gamma;\omega) \le 0.$$

Hence $\alpha(\gamma;\omega)$ is a monotonically decreasing function of γ for each real ω. The result follows on recalling (2.4) and noting that $\beta(\omega;s_0)$ is independent of γ. □

2.4 Relations to the LQG Cost

Our next aim is to relate entropy to the well-known concept of LQG cost [2, 38]. The LQG cost of a stable system is just the mean-square output of the system when it is driven by a white noise input. The formal definition is as follows.

Definition 2.4.1 (LQG cost) *Let $H \in \mathcal{RH}_\infty$, and let z be the output of this system when driven by a zero mean Gaussian white noise input w with spectrum equal to the identity. Then the associated* LQG cost *is defined by*

$$C(H) := \lim_{T\to\infty}\mathbf{E}\{W_T\},$$

where

$$W_T := \frac{1}{2T}\int_{-T}^{T} z^T(t)z(t)dt,$$

and where $\mathbf{E}$ *denotes expectation with respect to the noise.*

Remark 2.4.2 The above definition is actually of the *steady-state* LQG cost. That is, $C(H)$ is just the limit as $T \to \infty$ of the finite-time LQG cost

$$C_T(H) := \mathbf{E}\{W_T\}.$$

Remark 2.4.3 By Parseval's Theorem (see e.g., [65]), we have an equivalent frequency domain characterization [55, p107],[53]:

$$\begin{aligned} C(H) &= \frac{1}{2\pi}\int_{-\infty}^{\infty} \operatorname{trace}[H^*(j\omega)H(j\omega)]\, d\omega \\ &= \|H\|_2^2. \end{aligned}$$

Clearly $C(H)$ is finite if and only if $H(\infty) = 0$ (using a similar argument to that used to prove Proposition 2.3.1(iv)). This corresponds to none of the noise w appearing instantaneously at the output z.

For convenience, define

$$I(H;\infty;\infty) := \lim_{\gamma\to\infty}\{I(H;\gamma;\infty)\},$$

that is, the entropy evaluated when the $\mathcal{H}_\infty$-norm bound $\|H\|_\infty < \gamma$ is relaxed completely. We can now relate entropy and LQG cost, one of the main results of this chapter.

Theorem 2.4.4 *Let $H \in \mathcal{RH}_\infty$ with $H(\infty) = 0$. Let $\gamma \in \mathbb{R}$ be such that $\|H\|_\infty < \gamma$. Then*

(i) $I(H;\gamma;\infty) \geq C(H)$.

(ii) $I(H;\gamma;\infty) = C(H) + O(\gamma^{-2})$.

(iii) $I(H;\infty;\infty) = C(H)$.

Proof Appendix A.3. □

The vital observation is: for $\gamma < \infty$, the entropy is an *upper bound* on the LQG cost, but when the $\mathcal{H}_\infty$-norm bound is relaxed completely ($\gamma \to \infty$), the entropy is *equal* to the LQG cost. Part (ii) shows how fast the entropy converges to the LQG cost as γ increases.

2.5 The $\mathcal{H}_\infty$-Norm Bound

Here we review the interpretation of the $\mathcal{H}_\infty$-norm bound $\|H\|_\infty < \gamma$. Since the ideas are now well-known, we will be brief and refer the interested reader to [24] for full details.

There are two interpretations of the $\mathcal{H}_\infty$-norm which are of interest in a control systems context. The first stems from the fact [24] that the $\mathcal{H}_\infty$-norm is the induced norm from $\mathcal{RH}_2$ to $\mathcal{RH}_2$. In other words, if $H \in \mathcal{RH}_\infty$ and $w \in \mathcal{RH}_2$ then $Hw \in \mathcal{RH}_2$, and furthermore,

$$\|H\|_\infty = \sup_w\{\|Hw\|_2 \;:\; w \in \mathcal{RH}_2 \text{ and } \|w\|_2 \leq 1\}.$$

From an input-output point of view, this means that the maximum energy gain, from bounded energy inputs w to bounded energy outputs $z = Hw$, is equal to $\|H\|_\infty$. Hence $\|H\|_\infty < \gamma$ means that this energy gain is less than γ. If we interpret w as representing bounded energy disturbance signals, we see that $\|H\|_\infty < \gamma$ specifies a level of disturbance rejection.

The second interpretation of $\|H\|_\infty < \gamma$ is much more important, and is based on the well-known Small Gain Theorem. The version of the theorem of interest to us is stated below.

Theorem 2.5.1 (The Small Gain Theorem [66]) *Let $H \in \mathcal{RH}_\infty$. Suppose $\Delta \in \mathcal{RH}_\infty$ is connected from the output of H to the input of H, as illustrated in Figure 2.1. Then this closed-loop is (internally) stable if*

$$\sigma_1\{\Delta(j\omega)\}\,\sigma_1\{H(j\omega)\} < 1, \qquad \forall \omega \in \mathbb{R} \cup \{\infty\}.$$

Remark 2.5.2 In our setup, we have $H \in \mathcal{RH}_\infty$ and $\|H\|_\infty < \gamma$. Then by the Small Gain Theorem,

$$\Delta \in \mathcal{RH}_\infty \quad \text{and} \quad \|\Delta\|_\infty \leq \gamma^{-1}$$

is sufficient to guarantee stability of the system in Figure 2.1.

We can interpret Δ as a perturbation to the system, representing system uncertainty. Thus the Small Gain Theorem tells us that an $\mathcal{H}_\infty$-norm bound on H implies stability in the presence of $\mathcal{H}_\infty$-norm bounded system uncertainty. That is, the $\mathcal{H}_\infty$-norm bound implies a prespecified level of *stability robustness*. Notice that the size of tolerable uncertainty varies as γ^{-1}: robustness increases as the $\mathcal{H}_\infty$-norm bound on H decreases.

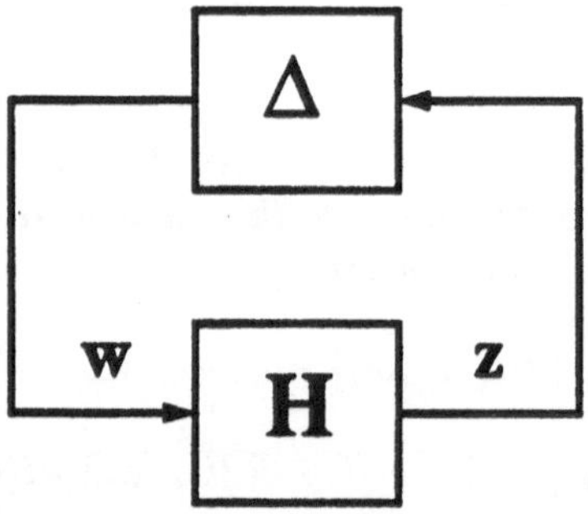

Figure 2.1: Block diagram for the Small Gain Theorem

Chapter 3

The Minimum Entropy $\mathcal{H}_\infty$ Control Problem

3.1 Introduction

In the previous chapter we saw how to define the entropy of an $\mathcal{H}_\infty$-norm bounded system. We also saw that a level of robustness is given by the $\mathcal{H}_\infty$-norm bound, and that the entropy overbounds the LQG cost. This analysis prompts us to pose the synthesis problem of finding a feedback controller which stabilizes a system subject to an $\mathcal{H}_\infty$-norm bound whilst simultaneously *minimizing* the entropy. We will solve this problem in this chapter. The solution exploits the parametrization of all stabilized closed-loop systems that meet the $\mathcal{H}_\infty$-norm bound. The central member of this set is shown to minimize the entropy at infinity and for this case a particularly simple formula is derived for the minimum value of the entropy.

3.2 Problem Formulation

Consider a system P with a state-space description

$$\begin{aligned} \dot{x} &= Ax + B_1 w + B_2 u \\ z &= C_1 x + D_{11} w + D_{12} u \\ y &= C_2 x + D_{21} w \end{aligned}$$

where the signals are as follows: $w \in \mathbb{R}^{m_1}$ is the disturbance vector, $u \in \mathbb{R}^{m_2}$ is the control input vector, $z \in \mathbb{R}^{p_1}$ is the error vector, $y \in \mathbb{R}^{p_2}$ is the observation vector and $x \in \mathbb{R}^n$ is the state vector. As is usual in $\mathcal{H}_\infty$ control problems we assume that $m_1 \geq p_2$, $p_1 \geq m_2$. The transfer function P from $[w^T \ \ u^T]^T$ to $[z^T \ \ y^T]^T$ is

$$P := \begin{bmatrix} P_{11} & P_{12} \\ P_{21} & P_{22} \end{bmatrix} = \left[\begin{array}{c|cc} A & B_1 & B_2 \\ \hline C_1 & D_{11} & D_{12} \\ C_2 & D_{21} & 0 \end{array}\right]$$

We connect a feedback controller K from y to u, as illustrated in Figure 3.1. This gives us a standard general framework into which a wide variety of control problems will fit (as explained in [24, Chapter 3]), so we shall adopt the usual convention of referring to P as a *standard plant*. This is to distinguish P from the actual plant, which is embedded in P together with the weighting functions and interconnections appropriate for the particular problem in hand. For example, a certain choice for P is made in Chapter 7—see equation (7.1). Until then, P is taken to be any standard plant, subject to the assumptions below.

The closed-loop transfer function H from disturbance vector w to error vector z is given by the (lower) linear fractional map

$$H = \mathcal{F}(P,K) := P_{11} + P_{12}K(I - P_{22}K)^{-1}P_{21}. \tag{3.1}$$

We need the following standard assumption which is necessary and sufficient for the existence of a controller K which stabilizes P.

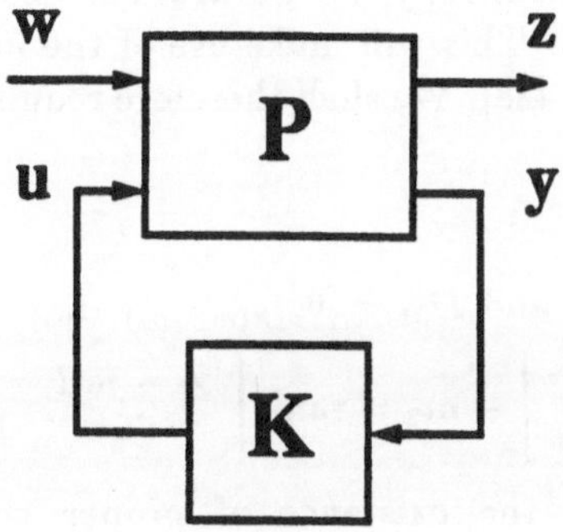

Figure 3.1: The closed-loop system

Assumption 3.2.1 (A, B_2) is stabilizable and (C_2, A) is detectable.

Remark 3.2.2 Note that by 'stabilizes' we mean 'internally stabilizes'. Also, K stabilizes P if and only if K stabilizes P_{22}. See [24, Chapter 4] for details.

The following definitions will be convenient in the sequel.

Definition 3.2.3 ((P,γ)-admissible controller) *Let P be a standard plant. A controller K is said to be* (P,γ)-admissible *if K stabilizes P and $\|\mathcal{F}(P,K)\|_\infty < \gamma$.*

Definition 3.2.4 ((P,γ)-admissible closed-loop) *A transfer function matrix $H \in \mathcal{RH}_\infty$ is said to be a* (P,γ)-admissible closed-loop *if $H = \mathcal{F}(P,K)$ and K is a (P,γ)-admissible controller.*

From the discussion in Section 2.5, the $\mathcal{H}_\infty$-norm bound γ provides robustness which increases as γ decreases. This motivates the usual $\mathcal{H}_\infty$-optimal control problem [24] of maximizing such robustness by minimizing the $\mathcal{H}_\infty$-norm bound γ.

Problem 3.2.5 (The $\mathcal{H}_\infty$-Optimal Control Problem)
Let P be a standard plant. Find a controller K_o and $\gamma_o \in \mathbb{R}$ which satisfy

$$\gamma_o = \inf_K \{\gamma \ : \ K \text{ is a } (P,\gamma)\text{-admissible controller}\} = \|\mathcal{F}(P,K_o)\|_\infty .$$

Pick $\gamma \in \mathbb{R}$ such that $\gamma > \gamma_o$. Then the set of (P,γ)-admissible controllers is non-empty. We can now impose our minimum entropy constraint.

Problem 3.2.6 (The Minimum Entropy $\mathcal{H}_\infty$ Control Problem) *Let P be a standard plant, let $\gamma > \gamma_o$, and let $s_0 \in (0, \infty]$. Minimize the closed-loop entropy $I(H; \gamma; s_0)$ over all (P, γ)-admissible closed-loops H.*

We will solve the Minimum Entropy $\mathcal{H}_\infty$ Control Problem by firstly parametrizing all (P, γ)-admissible closed-loops. This will make use of the parametrization of all (P, γ)-admissible controllers given in [28]. We shall therefore require the following assumptions to go with Assumption 3.2.1.

Assumptions 3.2.7

(i) $D_{12} = [0_{m_2 \times (p_1 - m_2)} \;\; I_{m_2}]^T$ and $D_{21} = [0_{p_2 \times (m_1 - p_2)} \;\; I_{p_2}]$.

(ii) $n = \operatorname{rank} \begin{bmatrix} A - j\omega I & B_2 \\ C_1 & D_{12} \end{bmatrix} - m_2 = \operatorname{rank} \begin{bmatrix} A - j\omega I & B_1 \\ C_2 & D_{21} \end{bmatrix} - p_2, \quad \forall \omega \in \mathbb{R}.$

Assumption 3.2.7(i) ensures the existence of proper controllers, whilst Assumption 3.2.7(ii) ensures [2] that the LQG problem associated with the plant P has an asymptotically stable closed-loop.

For completeness, we also state the LQG problem [2, 38] associated with the standard plant P. From Definition 2.4.1 and Remark 2.4.3 we see that the LQG cost is just $C(H) = \|H\|_2^2$, where $H = \mathcal{F}(P, K)$ is a stabilized closed-loop.

Problem 3.2.8 (The LQG Control Problem) *Let P be a standard plant. Minimize the LQG cost $C(\mathcal{F}(P, K))$ over all stabilized closed-loops $\mathcal{F}(P, K)$.*

The Minimum Entropy $\mathcal{H}_\infty$ Control Problem will be solved in Chapter 4, via the solution of the associated Minimum Entropy $\mathcal{H}_\infty$ Distance Problem, by exploiting recent state-space results of [30] on the latter problem. It is then possible to back-substitute to find the solution of the original Minimum Entropy $\mathcal{H}_\infty$ Control Problem. However, in this chapter, by exploiting the new results of [28, 29, 17], the Minimum Entropy $\mathcal{H}_\infty$ Control Problem is completely solved directly in terms of the original standard plant P without the intermediate step of a general distance problem. Furthermore, with entropy evaluated at infinity and assuming $D_{11} = 0$ (which guarantees that the minimum value of the entropy evaluated at infinity is finite) simple and explicit state-space formulae are given for the minimum entropy controller and the minimum value of the entropy. These are in terms of the state-space realization of P and the solutions to the two algebraic Riccati equations involved in parametrizing all (P, γ)-admissible controllers.

The remainder of this chapter is structured as follows. In the next section we derive the solution to the Minimum Entropy $\mathcal{H}_\infty$ Control Problem for arbitrary $s_0 \in (0, \infty)$, after firstly parametrizing all (P, γ)-admissible closed-loops. Section 3.4 deals with entropy at infinity; the minimum entropy solution is characterized in a particularly simple way in this case—explicit state-space formulae are derived in Section 3.5. Section 3.6 provides upper bounds on the $\mathcal{H}_\infty$-norm and LQG cost, and their inherent tradeoff, whilst Section 3.7 shows how to recover the LQG problem and its solution from the Minimum Entropy $\mathcal{H}_\infty$ Control Problem and its solution.

3.3 Solution in the General Case

We solve in this section the Minimum Entropy $\mathcal{H}_\infty$ Control Problem at an arbitrary $s_0 \in (0,\infty)$. To begin, all (P,γ)-admissible closed-loop transfer functions H are parametrized using the recent results of [28]. All (P,γ)-admissible closed-loops H are given as the linear fractional map of a stable *all-pass* transfer function matrix J_a and an arbitrary stable contraction Φ (i.e., $\Phi \in \mathcal{BRH}_\infty$). Then, by adapting the arguments of [4, 3], the choice of Φ which minimizes the entropy is derived, together with an expression for the resulting minimum value of the entropy. This problem has also been examined in [39], for the 'one-block' case when P_{12} and P_{21} are square.

To be specific, from [28], all (P,γ)-admissible controllers K can be written $K = \gamma\mathcal{F}(K_a,\Phi)$, where $\Phi \in \mathcal{BRH}_\infty$, and K_a can be calculated from the given data (i.e., P and γ). Throughout this chapter we find it convenient to keep the factor γ explicit in a slightly different way to [28]. Instead of using $K = \mathcal{F}(K_a,\Phi)$, with $\Phi \in \mathcal{RH}_\infty$ and $\|\Phi\|_\infty < \gamma$, we use $K = \gamma\mathcal{F}(K_a,\Phi)$ with $\Phi \in \mathcal{BRH}_\infty$, and redefine K_a accordingly. It follows that all (P,γ)-admissible closed-loops are given by

$$H = \mathcal{F}(P,\gamma\mathcal{F}(K_a,\Phi)) =: \gamma\mathcal{F}(J,\Phi).$$

It is then convenient to partition J into

$$J = \begin{bmatrix} J_{11} & J_{12} \\ J_{21} & J_{22} \end{bmatrix}$$

compatibly with Φ in $\mathcal{F}(J,\Phi)$. The definition of J is made clear in Figure 3.2.

Now J is asymptotically stable and contractive. It can therefore be thought of as a sub-block of an asymptotically stable and all-pass matrix J_a. The matrix J_a is an all-pass *dilation* of J. By performing the dilation of J into J_a we can establish the following lemma; details are given in [29].

Lemma 3.3.1 (All (P,γ)-admissible closed-loops)
All (P,γ)-admissible closed-loops H are given by

$$H = \gamma\mathcal{F}(J_a,\Theta)$$

where

$$\Theta := \begin{array}{c} m_2\updownarrow \\ p_1-m_2\updownarrow \end{array}\overset{\begin{array}{cc} \scriptstyle p_2 & \scriptstyle m_1-p_2 \end{array}}{\begin{bmatrix} \Phi & 0 \\ 0 & 0 \end{bmatrix}}, \qquad \Phi \in \mathcal{BRH}_\infty.$$

Furthermore, the matrix

$$J_a = \begin{array}{c} p_1\updownarrow \\ p_2\updownarrow \\ m_1-p_2\updownarrow \end{array}\overset{\begin{array}{ccc} \scriptstyle m_1 & \scriptstyle m_2 & \scriptstyle p_1-m_2 \end{array}}{\begin{bmatrix} J_{11} & J_{12} & J_{13} \\ J_{21} & J_{22} & J_{23} \\ J_{31} & J_{32} & J_{33} \end{bmatrix}}$$

has the following properties:

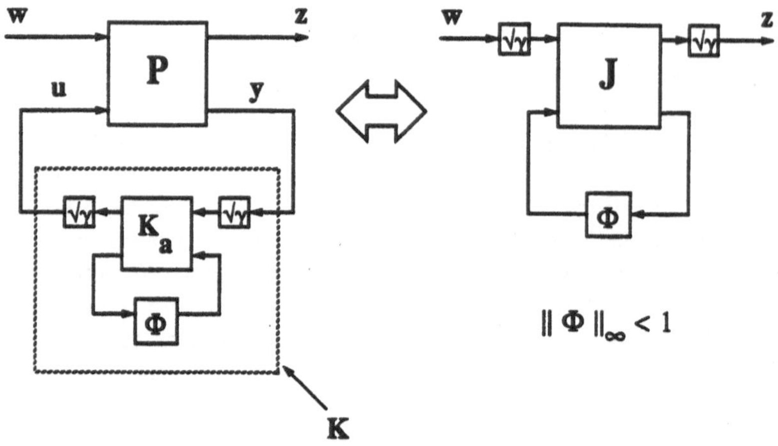

Figure 3.2: All closed-loops satisfying the $\mathcal{H}_\infty$-norm bound

(i) $J_a \in \mathcal{RH}_\infty$.

(ii) J_a *is all-pass.*

(iii) $J_{22}(\infty) = 0$.

It is now possible to exploit the structure of this parametrization in order to derive the choice of $\Phi \in \mathcal{BRH}_\infty$ which minimizes the closed-loop entropy at $s_0 \in (0, \infty)$.

Theorem 3.3.2
Consider the class of all (P, γ)-admissible closed-loops H, as parametrized in Lemma 3.3.1. Then $I(H; \gamma; s_0)$, the closed-loop entropy at $s_0 \in (0, \infty)$, is minimized over this class by the unique choice

$$\Phi = J_{22}^*(s_0).$$

Proof Here we adapt the method of [4, 3]. From Lemma 3.3.1, we seek a $\Phi \in \mathcal{BRH}_\infty$ which minimizes $I(H; \gamma; s_0)$, where all (P, γ)-admissible closed-loops H are

$$H = \gamma \mathcal{F}(J_a, \begin{bmatrix} \Phi & 0 \\ 0 & 0 \end{bmatrix}).$$

Now J_a is all-pass so $J_a^* J_a = I$, which can be expressed in terms of its block partitions as

$$I = J_{11}^* J_{11} + \begin{bmatrix} J_{21} \\ J_{31} \end{bmatrix}^* \begin{bmatrix} J_{21} \\ J_{31} \end{bmatrix} \tag{3.2}$$

$$0 = J_{11}^* \begin{bmatrix} J_{12} & J_{13} \end{bmatrix} + \begin{bmatrix} J_{21} \\ J_{31} \end{bmatrix}^* \begin{bmatrix} J_{22} & J_{23} \\ J_{32} & J_{33} \end{bmatrix} \tag{3.3}$$

$$I = \begin{bmatrix} J_{12} & J_{13} \end{bmatrix}^* \begin{bmatrix} J_{12} & J_{13} \end{bmatrix} + \begin{bmatrix} J_{22} & J_{23} \\ J_{32} & J_{33} \end{bmatrix}^* \begin{bmatrix} J_{22} & J_{23} \\ J_{32} & J_{33} \end{bmatrix} \tag{3.4}$$

By exploiting (3.2), (3.3), (3.3)*, and (3.4), it may be shown by straightforward manipulations that

$$\begin{aligned} I - \gamma^{-2} H^* H &= \begin{bmatrix} J_{21} \\ J_{31} \end{bmatrix}^* \left\{ \begin{bmatrix} I & 0 \\ 0 & I \end{bmatrix} - \begin{bmatrix} J_{22} & J_{23} \\ J_{32} & J_{33} \end{bmatrix} \begin{bmatrix} \Phi & 0 \\ 0 & 0 \end{bmatrix} \right\}^{-*} \\ &\quad \times \begin{bmatrix} I - \Phi^*\Phi & 0 \\ 0 & I \end{bmatrix} \left\{ \begin{bmatrix} I & 0 \\ 0 & I \end{bmatrix} - \begin{bmatrix} J_{22} & J_{23} \\ J_{32} & J_{33} \end{bmatrix} \begin{bmatrix} \Phi & 0 \\ 0 & 0 \end{bmatrix} \right\}^{-1} \begin{bmatrix} J_{21} \\ J_{31} \end{bmatrix} \end{aligned}$$

Upon substituting this into the definition of $I(H;\gamma;s_0)$ (see Definition 2.2.1) it follows that

$$\begin{aligned} I(H;\gamma;s_0) &= \gamma^2 I(\Phi;1;s_0) \\ &\quad - \frac{\gamma^2}{2\pi} \int_{-\infty}^{\infty} \ln \left| \det \begin{bmatrix} J_{21}(j\omega) \\ J_{31}(j\omega) \end{bmatrix}^* \begin{bmatrix} J_{21}(j\omega) \\ J_{31}(j\omega) \end{bmatrix} \right| \left[\frac{s_0^2}{s_0^2 + \omega^2} \right] d\omega \\ &\quad + \frac{\gamma^2}{\pi} \int_{-\infty}^{\infty} \ln |\det(I - J_{22}(j\omega)\Phi(j\omega))| \left[\frac{s_0^2}{s_0^2 + \omega^2} \right] d\omega \\ &=: \gamma^2 I(\Phi;1;s_0) \quad + \quad (a) \quad + \quad (b). \end{aligned}$$

Equation (3.2) allows us to write

$$(a) = \gamma^2 I(J_{11};1;s_0).$$

By Lemma 3.3.1, both Φ and J_{22} are in $\mathcal{RH}_\infty$. Furthermore, $\|\Phi\|_\infty < 1$ and $\|J_{22}\|_\infty \leq 1$ (since J_{22} is part of the all-pass matrix J_a). This is enough to guarantee that $\det(I - J_{22}\Phi)^{\pm 1} \in \mathcal{RH}_\infty$, allowing the use of Poisson's Integral Theorem [52, Theorem 17.16] to evaluate (b):

$$(b) = \gamma^2 s_0 \ln |\det(I - J_{22}(s_0)\Phi(s_0))|.$$

Therefore

$$I(H;\gamma;s_0) = \gamma^2 I(\Phi;1;s_0) + \gamma^2 I(J_{11};1;s_0) + \gamma^2 s_0 \ln |\det(I - J_{22}(s_0)\Phi(s_0))|. \tag{3.5}$$

To proceed, we reparametrize as follows. Define a constant Julia matrix [65, p148] by

$$L = \begin{bmatrix} L_{11} & L_{12} \\ L_{21} & L_{22} \end{bmatrix} := \begin{bmatrix} -J_{22}^*(s_0) & (I - J_{22}^*(s_0)J_{22}(s_0))^{1/2} \\ (I - J_{22}(s_0)J_{22}^*(s_0))^{1/2} & J_{22}(s_0) \end{bmatrix}.$$

It is easy to verify that L is unitary. Then define a new arbitrary stable parameter $\tilde{\Phi} \in \mathcal{BRH}_\infty$, by

$$\tilde{\Phi} = \mathcal{F}(L, \Phi), \qquad \Phi \in \mathcal{BRH}_\infty.$$

Note that this maps $\mathcal{BRH}_\infty$ onto itself, and in particular

$$\tilde{\Phi} = 0 \iff \Phi = J_{22}^*(s_0). \tag{3.6}$$

By applying an identical argument to that used to obtain (3.5) above, it can be shown that

$$I(\tilde{\Phi}; 1; s_0) = I(\Phi; 1; s_0) + I(J_{22}^*(s_0); 1; s_0) + s_0 \ln |\det(I - J_{22}(s_0)\Phi(s_0))|. \tag{3.7}$$

Eliminating Φ between (3.5) and (3.7) gives

$$I(H; \gamma; s_0) = \gamma^2 I(\tilde{\Phi}; 1; s_0) + \gamma^2 I(J_{11}; 1; s_0) - \gamma^2 I(J_{22}^*(s_0); 1; s_0). \tag{3.8}$$

It is immediate from (3.8) that $I(H; \gamma; s_0)$ is minimized by the unique choice $\tilde{\Phi} = 0$. To complete the proof, it is only necessary to recall (3.6). □

Denote 'minimum entropy at s_0' quantities by $(\,\cdot\,)_{ME_{s_0}}$. The following corollary is then immediate.

Corollary 3.3.3 *The minimum value of $I(H; \gamma; s_0)$, as H varies over all (P, γ)-admissible closed-loops, is*

$$I(H_{ME_{s_0}}; \gamma; s_0) = \gamma^2 I(J_{11}; 1; s_0) - \gamma^2 I(J_{22}^*(s_0); 1; s_0).$$

Proof Put $\tilde{\Phi} = 0$ in (3.8). □

3.4 Solution at Infinity

In this section we concentrate on the special case of the Minimum Entropy $\mathcal{H}_\infty$ Control Problem based on $I(H; \gamma; \infty)$, the *entropy at infinity*. As remarked in Chapter 2, this is the most interesting and important case because of the strong connections with other control problems. See [20] for entropy at infinity in the context of Nehari interpolation.

Remark 3.4.1 From Proposition 2.3.1(iv), $I(H; \gamma; \infty)$ is finite if and only if $H(\infty) = 0$. Since $H(\infty) = D_{11} + D_{12}K(\infty)D_{21}$, a necessary and sufficient condition for $H(\infty) = 0$ is that $[I_{p_1 - m_2} \;\; 0_{(p_1-m_2)\times m_2}]D_{11} = 0$ and $D_{11}[I_{m_1-p_2} \;\; 0_{(m_1-p_2)\times p_2}]^T = 0$. In this case an equivalent problem with $D_{11} = 0$ is easily derived by making the change of control input variable from u to v, given by $v = u + D_{12}^T D_{11} D_{21}^T y$.

It is convenient to introduce the notation

$$H\{\Phi\} := \gamma \mathcal{F}(J_a, \begin{bmatrix} \Phi & 0 \\ 0 & 0 \end{bmatrix})$$

to denote a (P,γ)-admissible closed-loop produced via Lemma 3.3.1 by a $\Phi \in \mathcal{BRH}_\infty$. Also define

$$\begin{aligned}\Phi_{ME\infty} &:= \lim_{s_0\to\infty}\{\Phi_{MEs_0}\}\\ &= \lim_{s_0\to\infty}\{J_{22}^*(s_0)\}\\ &= 0,\end{aligned}$$

where the second equality is taken from Theorem 3.3.2 and the third equality is from Lemma 3.3.1(iii). Then

$$H\{\Phi_{ME\infty}\} = H\{0\} = \gamma J_{11}. \tag{3.9}$$

The minimum entropy at infinity solution is obtained by taking the limit, as $s_0 \to \infty$, of the minimum entropy solution at finite s_0. A little care is needed when the minimum value of the entropy at infinity is not finite. This is all formalized in the next theorem.

Theorem 3.4.2

(i) *The choice*

$$\Phi = \Phi_{ME\infty} = 0$$

in the parametrization of Lemma 3.3.1 of all (P,γ)-admissible closed-loops H, is the unique choice of Φ which satisfies:

$$\begin{aligned}&\forall\Psi \in \mathcal{BRH}_\infty \quad (\Psi \neq 0):\\ &\quad \exists s(\Psi) < \infty \quad \textit{such that} \quad \forall s_0 > s(\Psi): \quad I(H\{\Phi_{ME\infty}\};\gamma;s_0) < I(H\{\Psi\};\gamma;s_0).\end{aligned}$$

(ii) *If $J_{11}(\infty) = 0$ then $\Phi = \Phi_{ME\infty} = 0$ is the unique choice of Φ which minimizes $I(H\{\Phi\};\gamma;\infty)$ over $\Phi \in \mathcal{BRH}_\infty$.*

Proof *Part (i)* In view of (3.9), put $H\{\Phi_{ME\infty}\} = \gamma J_{11}$ into equation (3.5). Then for any $\Psi \in \mathcal{BRH}_\infty$,

$$\begin{aligned}I(H\{\Phi_{ME\infty}\};\gamma;s_0) - I(H\{\Psi\};\gamma;s_0) = &\ -\gamma^2 I(\Psi;1;s_0)\\ &\ -\gamma^2 s_0 \ln|\det(I - J_{22}(s_0)\Psi(s_0))|\\ =: &\ \ a(s_0) \ + \ b(s_0).\end{aligned}$$

Consider $\Psi \in \mathcal{BRH}_\infty$ such that $\Psi \neq 0$. Then $I(\Psi;1;s_0) \neq 0$, and it is clear that $a(s_0) < 0$ and $b(s_0) \geq 0$ for all $s_0 \in (0,\infty]$. If Ψ is proper, but not strictly proper, then $a(s_0) \to -\infty$ (monotonically for each fixed Ψ, by the argument in the proof of Proposition 2.3.1(iv)) as $s_0 \to \infty$. An application of Lemma A.2.1(i) of Appendix A, shows that $b(s_0) \to b$, a constant, as $s_0 \to \infty$. This guarantees the existence of a positive number $s(\Psi) < \infty$, such that $a(s_0)+b(s_0) < 0$ for all $s_0 > s(\Psi)$. Alternatively, if Ψ is strictly proper, then $a(s_0) \to a < 0$, a constant, whereas (by Lemma A.2.1(i) again) $b(s_0) \to 0$ as $s_0 \to \infty$, and once more there exists a positive number $s(\Psi) < \infty$, such that $a(s_0) + b(s_0) < 0$ for all $s_0 > s(\Psi)$. This establishes that $\Phi_{ME\infty} = 0$ does have the property stated in the Part (i) of the theorem.

Note that the above also proves that $\Phi_{ME\infty} = 0$ gives a strictly smaller entropy than any other $\Psi \in \mathcal{BRH}_\infty$ for sufficiently large s_0, hence this $\Phi_{ME\infty} = 0$ is unique, completing the proof of Part (i).

Part (ii) By hypothesis, $J_{11}(\infty) = 0$, which implies that the minimum entropy closed-loop, $H\{\Phi_{ME\infty}\} = \gamma J_{11}$, is strictly proper. By Proposition 2.3.1(iv), the minimum value of $I(H;\gamma;\infty)$ is then finite.

Suppose the claim of the Part (ii) is false, i.e., that there exists a $\Phi \in \mathcal{BRH}_\infty$, call it Φ_1, such that

$$I(H\{\Phi_{ME\infty}\};\gamma;\infty) - I(H\{\Phi_1\};\gamma;\infty) = c, \quad \text{for some } c > 0. \tag{3.10}$$

Obviously this is only possible if $H\{\Phi_1\}$ is strictly proper, so that $I(H\{\Phi_1\};\gamma;\infty)$ is finite, and there exists $s_1 \in (0,\infty)$, such that

$$|I(H\{\Phi_1\};\gamma;\infty) - I(H\{\Phi_1\};\gamma;s_0)| < c/6, \quad \forall s_0 > s_1. \tag{3.11}$$

Similarly, there exists $s_2 \in (0,\infty)$, such that

$$|I(H\{\Phi_{ME\infty}\};\gamma;s_0) - I(H\{\Phi_{ME\infty}\};\gamma;\infty)| < c/6, \quad \forall s_0 > s_2. \tag{3.12}$$

The entropy evaluation in (3.5), together with $\Phi_{ME\infty} = 0$ and $\Phi_{ME s_0} = J_{22}^*(s_0)$, gives

$$\begin{aligned} I(H\{\Phi_{ME s_0}\};\gamma;s_0) - I(H\{\Phi_{ME\infty}\};\gamma;s_0) &= -\gamma^2 s_0 \ln|\det(I - J_{22}(s_0)J_{22}^*(s_0))| \\ &\to 0 \text{ as } s_0 \to \infty \end{aligned}$$

by Lemma A.2.1(i) of Appendix A, since $J_{22}(s_0) = O(s_0^{-1})$. Thus, there exists $s_3 \in (0,\infty)$, such that

$$|I(H\{\Phi_{ME s_0}\};\gamma;s_0) - I(H\{\Phi_{ME\infty}\};\gamma;s_0)| < c/6, \quad \forall s_0 > s_3. \tag{3.13}$$

Using (3.10)-(3.13) in conjunction with the identity

$$\begin{aligned} I(H\{\Phi_{ME s_0}\};\gamma;s_0) - I(H\{\Phi_1\};\gamma;s_0) = {} & I(H\{\Phi_{ME s_0}\};\gamma;s_0) - I(H\{\Phi_{ME\infty}\};\gamma;s_0) \\ & + I(H\{\Phi_{ME\infty}\};\gamma;s_0) - I(H\{\Phi_{ME\infty}\};\gamma;\infty) \\ & + I(H\{\Phi_{ME\infty}\};\gamma;\infty) - I(H\{\Phi_1\};\gamma;\infty) \\ & + I(H\{\Phi_1\};\gamma;\infty) - I(H\{\Phi_1\};\gamma;s_0) \end{aligned}$$

gives, for all $s_0 > \max\{s_1, s_2, s_3\}$,

$$I(H\{\Phi_{ME s_0}\};\gamma;s_0) - I(H\{\Phi_1\};\gamma;s_0) > -c/6 + c - c/6 - c/6 = c/2 > 0.$$

So for all sufficiently large s_0,

$$I(H\{\Phi_{ME s_0}\};\gamma;s_0) - I(H\{\Phi_1\};\gamma;s_0) > 0,$$

which contradicts the fact that $\Phi_{ME s_0}$ is the minimum entropy solution at s_0. Therefore, (3.10) cannot hold, which establishes that $\Phi_{ME\infty} = 0$ does indeed solve the Minimum Entropy $\mathcal{H}_\infty$ Control Problem at infinity. The proof is completed by noting that uniqueness of this $\Phi_{ME\infty}$ is implied by the proof of uniqueness of $\Phi_{ME\infty}$ in Part (i). □

The minimum entropy at infinity solution therefore has the simple and appealing characterization of $\Phi = 0$: it is the 'central solution' from the ball of admissible closed-loops. The following corollary arises by using this important fact in Lemma 3.3.1, Corollary 3.3.3 and in the controller given in [28].

Corollary 3.4.3 *With entropy evaluated at infinity, the unique controller which solves the Minimum Entropy $\mathcal{H}_\infty$ Control Problem (Problem 3.2.6) is*

$$K_{ME\infty} = \gamma\mathcal{F}(K_a, 0) = \gamma(K_a)_{11},$$

that is, the central (P,γ)-admissible controller.

The minimum entropy closed-loop is

$$H_{ME\infty} = \mathcal{F}(P, K_{ME\infty}) = \gamma\mathcal{F}(J_a, 0) = \gamma J_{11},$$

that is, the central (P,γ)-admissible closed-loop.

The minimum value of the closed-loop entropy is

$$I(H_{ME\infty}; \gamma; \infty) = \gamma^2 I(J_{11}; 1; \infty).$$

3.5 State-Space Formulae for the Minimum Entropy

In this section we derive an explicit state-space solution to the Minimum Entropy $\mathcal{H}_\infty$ Control Problem with entropy evaluated at infinity. In particular, we shall derive simple formulae for the minimum value of the entropy. For the remainder of this section we assume $D_{11} = 0$. Then the minimum value of the entropy at infinity is finite (Remark 3.4.1), and is given by the state-space formulae of the following theorem.

Theorem 3.5.1 *Assume that $D_{11} = 0$. Recall the standing assumptions that $D_{12} = [0_{m_2\times(p_1-m_2)}\ \ I_{m_2}]^T$ and that $D_{21} = [0_{p_2\times(m_1-p_2)}\ \ I_{p_2}]$. Let $D_\perp$ make $[D_{12}\ \ D_\perp]$ square and orthogonal, and let $\tilde{D}_\perp$ make $[D_{21}^T\ \ \tilde{D}_\perp^T]^T$ square and orthogonal. Let $X_\infty = X_\infty^T \geq 0$ be the solution to the algebraic Riccati equation*

$$\begin{aligned}0 = X_\infty(A - B_2D_{12}^TC_1) + (A - B_2D_{12}^TC_1)^TX_\infty \\ + C_1^TD_\perp D_\perp^TC_1 + X_\infty(\gamma^{-2}B_1B_1^T - B_2B_2^T)X_\infty \qquad (3.14)\end{aligned}$$

such that

$$A - B_2D_{12}^TC_1 + (\gamma^{-2}B_1B_1^T - B_2B_2^T)X_\infty \quad \textit{is asymptotically stable,}$$

and let $Y_\infty = Y_\infty^T \geq 0$ be the solution to the algebraic Riccati equation

$$\begin{aligned} 0 = Y_\infty(A - B_1 D_{21}^T C_2)^T &+ (A - B_1 D_{21}^T C_2) Y_\infty \\ &+ B_1 \tilde{D}_\perp^T \tilde{D}_\perp B_1^T + Y_\infty(\gamma^{-2} C_1^T C_1 - C_2^T C_2) Y_\infty \end{aligned} \tag{3.15}$$

such that

$$A - B_1 D_{21}^T C_2 + Y_\infty(\gamma^{-2} C_1^T C_1 - C_2^T C_2) \quad \textit{is asymptotically stable.}$$

Define

$$Z_\infty := (I - \gamma^{-2} Y_\infty X_\infty)^{-1}.$$

Then the minimum value of the closed-loop entropy at infinity is

$$\begin{aligned} I(H_{ME\infty}; \gamma; \infty) &= \operatorname{trace}[C_1 Y_\infty C_1^T] \\ &\quad + \operatorname{trace}[(D_{21} B_1^T + C_2 Y_\infty) X_\infty Z_\infty (B_1 D_{21}^T + Y_\infty C_2^T)] \quad (3.16) \\ &= \operatorname{trace}[B_1^T X_\infty B_1] \\ &\quad + \operatorname{trace}[(D_{12}^T C_1 + B_2^T X_\infty) Z_\infty Y_\infty (C_1^T D_{12} + X_\infty B_2)]. \quad (3.17) \end{aligned}$$

Remark 3.5.2 An alternative proof to the one given in the remainder of this section is given in Appendix B. The derivation there is based on the approaches developed in [17, 28, 29]. It has the advantage of providing an interpretation of the separate terms making up the total entropy. To use the terminology from [17, 29], it is shown in Appendix B that the total entropy in (3.17) is composed of the 'Full Information' entropy plus the entropy of the derived 'Output Estimation' problem. Alternatively, the total entropy in (3.16) is composed of the 'Full Control' entropy plus the entropy of the derived 'Disturbance Feedforward' problem.

Remark 3.5.3 Note that from [28] we know that there exists $X_\infty \geq 0$ and $Y_\infty \geq 0$ as in the theorem, with $\rho(X_\infty Y_\infty) < \gamma^2$, if and only if $\gamma > \gamma_o$.

Remark 3.5.4 It is valid to refer to X_∞ (or Y_∞) as *the* stabilizing solution to its Riccati equation because [36, Lemma 3.4-1] if a Riccati equation has a stabilizing solution then it is unique.

Proof of Theorem 3.5.1 The remainder of this section will be devoted to proving Theorem 3.5.1. We will prove formula (3.16) in detail. The dual formula (3.17) may be proved in a similar way, based on the fact (Proposition 2.3.1(vii)) that the entropy of H is the same as the entropy of H^T. Since the proof is quite long, it is split into a number of intermediate lemmas.

The next lemma gives state-space formulae for the all-pass matrix J_a needed in Lemma 3.3.1 to generate all (P, γ)-admissible closed-loops.

Lemma 3.5.5 *With assumptions and definitions as in Theorem 3.5.1, the all-pass matrix J_a of Lemma 3.3.1 has a state-space realization*

$$J_a = \left[\begin{array}{cc|ccc} A & B_2\hat{C}_1 & \gamma^{-1/2}B_1 & \gamma^{1/2}B_2 & \gamma^{1/2}B_3 \\ \hat{B}_1C_2 & \hat{A} & \gamma^{-1/2}\hat{B}_1D_{21} & \gamma^{1/2}\hat{B}_2 & \gamma^{1/2}\hat{B}_3 \\ \hline \gamma^{-1/2}C_1 & \gamma^{-1/2}D_{12}\hat{C}_1 & 0 & D_{12} & D_\perp \\ \gamma^{1/2}C_2 & \gamma^{1/2}\hat{C}_2 & D_{21} & 0 & 0 \\ \gamma^{1/2}C_3 & \gamma^{1/2}\hat{C}_3 & \tilde{D}_\perp & 0 & 0 \end{array}\right]$$

where

$$\begin{array}{ll} \hat{B}_1 = B_1D_{21}^T + Y_\infty C_2^T & \hat{C}_1 = -(D_{12}^TC_1 + B_2^TX_\infty)Z_\infty \\ \hat{B}_2 = B_2 + \gamma^{-2}Y_\infty C_1^TD_{12} & \hat{C}_2 = -(C_2 + \gamma^{-2}D_{21}B_1^TX_\infty)Z_\infty \\ \hat{B}_3 = -Z_\infty^{-1}X_\infty^\dagger C_1^TD_\perp & \hat{C}_3 = \tilde{D}_\perp B_1^TY_\infty^\dagger \\ B_3 = -X_\infty^\dagger C_1^TD_\perp & C_3 = -\tilde{D}_\perp B_1^TY_\infty^\dagger \end{array}$$

and

$$\begin{aligned} \hat{A} &= A - B_1D_{21}^TC_2 + Y_\infty(\gamma^{-2}C_1^TC_1 - C_2^TC_2) + \hat{B}_2\hat{C}_1 \\ &= Z_\infty^{-1}(A - B_2D_{12}^TC_1 + (\gamma^{-2}B_1B_1^T - B_2B_2^T)X_\infty)Z_\infty + \hat{B}_1\hat{C}_2. \end{aligned}$$

Proof See Appendix B. □

Remark 3.5.6 Having already established that $\Phi_{ME\infty} = 0$, we immediately have from the above lemma together with Corollary 3.4.3 and [28] that, in the case of entropy evaluated at infinity and $D_{11} = 0$, the unique minimum entropy $\mathcal{H}_\infty$ controller has a realization

$$\begin{aligned} K_{ME\infty} = \gamma(K_a)_{11} &= \left[\begin{array}{c|c} \hat{A} & \hat{B}_1 \\ \hline \hat{C}_1 & 0 \end{array}\right] \\ &= \left[\begin{array}{c|c} \hat{A} & B_1D_{21}^T + Y_\infty C_2^T \\ \hline -(D_{12}^TC_1 + B_2^TX_\infty)Z_\infty & 0 \end{array}\right]. \end{aligned}$$

This is an n-state realization, as is the given realization of P. The minimum entropy closed-loop has a realization

$$\begin{aligned} H_{ME\infty} &= \gamma J_{11} \\ &= \left[\begin{array}{cc|c} A & -B_2(D_{12}^TC_1 + B_2^TX_\infty)Z_\infty & B_1 \\ (B_1D_{21}^T + Y_\infty C_2^T)C_2 & \hat{A} & (B_1D_{21}^T + Y_\infty C_2^T)D_{21} \\ \hline C_1 & -D_{12}(D_{12}^TC_1 + B_2^TX_\infty)Z_\infty & 0 \end{array}\right] \end{aligned}$$

Note that the above formulae for the minimum entropy $\mathcal{H}_\infty$ controller, closed-loop *and* the resulting entropy only require the solutions of two independent algebraic Riccati equations, one for X_∞ and one for Y_∞.

We now perform a state-transformation which will make later calculations clearer.

Lemma 3.5.7 *There exists a nonsingular state-transformation S which, when applied to the standard plant P, leads to*

$$Y_\infty = \begin{bmatrix} Y_1 & 0 \\ 0 & 0 \end{bmatrix}, \qquad Y_1 > 0.$$

Then for

$$N := A - B_1 D_{21}^T C_2,$$

partitioned compatibly with Y_∞ as follows,

$$N = \begin{bmatrix} N_{11} & N_{12} \\ N_{21} & N_{22} \end{bmatrix}, \quad \begin{bmatrix} C_1 \\ C_2 \end{bmatrix} = \begin{bmatrix} C_{11} & C_{12} \\ C_{21} & C_{22} \end{bmatrix}, \quad B_1 = \begin{bmatrix} B_{11} \\ B_{12} \end{bmatrix},$$

we have that

(i) $B_{12}\tilde{D}_\perp^T = 0.$

(ii) $N_{21} = 0.$

(iii) $Y_1 N_{11}^T + N_{11} Y_1 + Y_1(\gamma^{-2} C_{11}^T C_{11} - C_{21}^T C_{21}) Y_1 + B_{11}\tilde{D}_\perp^T \tilde{D}_\perp B_{11}^T = 0.$

(iv) $N_{11} + Y_1(\gamma^{-2} C_{11}^T C_{11} - C_{21}^T C_{21})$ *is asymptotically stable.*

(v) N_{22} *is asymptotically stable.*

Proof Since under a nonsingular state-transformation S, $Y_\infty \to S Y_\infty S^T$, and $Y_\infty \geq 0$, such an S clearly exists. Upon rewriting the Y_∞ Riccati equation (3.15) in the implied block form, Part (i) is obtained immediately from the (2,2) block; Part (ii) follows from the (2,1) block and Part (i); Part (iii) is just the (1,1) block. Furthermore, it follows that

$$\begin{aligned} &A - B_1 D_{21}^T C_2 + Y_\infty(\gamma^{-2} C_1^T C_1 - C_2^T C_2) \\ &\quad = \begin{bmatrix} N_{11} + Y_1(\gamma^{-2} C_{11}^T C_{11} - C_{21}^T C_{21}) & N_{12} + Y_1(\gamma^{-2} C_{11}^T C_{12} - C_{21}^T C_{22}) \\ 0 & N_{22} \end{bmatrix} \end{aligned}$$

and since from Lemma 3.5.5, $A - B_1 D_{21}^T C_2 + Y_\infty(\gamma^{-2} C_1^T C_1 - C_2^T C_2)$ is asymptotically stable, we conclude that both $N_{11} + Y_1(\gamma^{-2} C_{11}^T C_{11} - C_{21}^T C_{21})$ and N_{22} are too, which proves Parts (iv) and (v). □

At this point we recall from Corollary 3.4.3 that the minimum value of the entropy at infinity is just

$$\begin{aligned} I(H_{ME\infty}; \gamma; \infty) &= \gamma^2 I(J_{11}; 1; \infty) \\ &= \lim_{s_0 \to \infty} \left\{ -\frac{\gamma^2}{2\pi} \int_{-\infty}^{\infty} \ln \left| \det \begin{bmatrix} J_{21}(j\omega) \\ J_{31}(j\omega) \end{bmatrix}^* \right. \right. \\ &\qquad \left. \left. \times \begin{bmatrix} J_{21}(j\omega) \\ J_{31}(j\omega) \end{bmatrix} \right| \left[\frac{s_0^2}{s_0^2 + \omega^2} \right] d\omega \right\} \end{aligned} \tag{3.18}$$

where (3.2) has been used to get (3.18). From Lemma 3.5.5,

$$\begin{bmatrix} J_{21} \\ J_{31} \end{bmatrix} = \left[\begin{array}{cc|c} A & B_2\hat{C}_1 & B_1 \\ \hat{B}_1C_2 & \hat{A} & \hat{B}_1D_{21} \\ \hline C_2 & \hat{C}_2 & D_{21} \\ C_3 & \hat{C}_3 & \tilde{D}_\perp \end{array}\right] =: \left[\begin{array}{c|c} \bar{A} & \bar{B} \\ \hline \bar{C} & \bar{D} \end{array}\right]. \tag{3.19}$$

Note that

$$\begin{bmatrix} J_{21} \\ J_{31} \end{bmatrix} \in \mathcal{RH}_\infty$$

because it is part of the asymptotically stable all-pass matrix J_a, but its inverse is not in $\mathcal{RH}_\infty$. Lemma A.4.2 (Appendix A) shows us how to evaluate the integral in (3.18) for such a transfer function matrix: we find that

$$I(H_{ME\infty};\gamma;\infty) = -\gamma^2 \operatorname{trace}[\bar{D}^{-1}\bar{C}\bar{B}] - 2\gamma^2 \sum_{\operatorname{Re}\{\lambda_i\}>0} \lambda_i\{\bar{A} - \bar{B}\bar{D}^{-1}\bar{C}\} \tag{3.20}$$

To evaluate the second term we need to evaluate the unstable eigenvalues of $\bar{A}-\bar{B}\bar{D}^{-1}\bar{C}$, the '$A$' matrix of $\begin{bmatrix} J_{21} \\ J_{31} \end{bmatrix}^{-1}$. This is done in the next lemma.

Lemma 3.5.8 *With notation as above,*

$$2\sum_{\operatorname{Re}\{\lambda_i\}>0} \lambda_i\{\bar{A} - \bar{B}\bar{D}^{-1}\bar{C}\} = \operatorname{trace}[B_1\tilde{D}_\perp^T\tilde{D}_\perp B_1^T Y_\infty^\dagger - Y_\infty(\gamma^{-2}C_1^TC_1 - C_2^TC_2)].$$

Proof Denoting a 'don't care' element of a matrix by $(*)$, it is straightforward to exploit Lemmas 3.5.5 and 3.5.7 to show that

$$\begin{aligned} \bar{A} - \bar{B}\bar{D}^{-1}\bar{C} &= \begin{bmatrix} A - B_1D_{21}^TC_2 - B_1\tilde{D}_\perp^TC_3 & (*) \\ \hat{B}_1C_2 - \hat{B}_1C_2 - \hat{B}_1D_{21}\tilde{D}_\perp^TC_3 & \hat{A} - \hat{B}_1\hat{C}_2 - \hat{B}_1D_{21}\tilde{D}_\perp^T\hat{C}_3 \end{bmatrix} \\ &= \begin{bmatrix} A - B_1D_{21}^TC_2 + B_1\tilde{D}_\perp^T\tilde{D}_\perp B_1^TY_\infty^\dagger & (*) \\ 0 & Z_\infty^{-1}MZ_\infty \end{bmatrix} \\ &= \begin{bmatrix} -Y_1N_{11}^TY_1^{-1} - Y_1(\gamma^{-2}C_{11}^TC_{11} - C_{21}^TC_{21}) & (*) & (*) \\ 0 & N_{22} & (*) \\ 0 & 0 & Z_\infty^{-1}MZ_\infty \end{bmatrix} \end{aligned}$$

where $M := A - B_2D_{12}^TC_1 + (\gamma^{-2}B_1B_1^T - B_2B_2^T)X_\infty$. Therefore,

$$\begin{aligned} \left\{\lambda_i\{\bar{A} - \bar{B}\bar{D}^{-1}\bar{C}\}\right\} &= \left\{-\lambda_i\{N_{11} + Y_1(\gamma^{-2}C_{11}^TC_{11} - C_{21}^TC_{21})\}\right\} \\ &\quad \cup \{\lambda_i\{N_{22}\}\} \cup \{\lambda_i\{M\}\}. \end{aligned}$$

By Lemma 3.5.7(v), N_{22} is asymptotically stable, and by Lemma 3.5.5, so is M. Furthermore, by Lemma 3.5.7(iv), $-(N_{11} + Y_1(\gamma^{-2}C_{11}^TC_{11} - C_{21}^TC_{21}))$ is antistable. Hence

$$\begin{aligned} \sum_{\operatorname{Re}\{\lambda_i\}>0} \lambda_i\{\bar{A} - \bar{B}\bar{D}^{-1}\bar{C}\} &= -\sum_i \lambda_i\{N_{11} + Y_1(\gamma^{-2}C_{11}^TC_{11} - C_{21}^TC_{21})\} \\ &= \operatorname{trace}[Y_1N_{11}^TY_1^{-1} + B_{11}\tilde{D}_\perp^T\tilde{D}_\perp B_{11}^TY_1^{-1}] \end{aligned}$$

(using Lemma 3.5.7(iii)Y_1^{-1}) and therefore, using well-known properties of trace[·],

$$\begin{aligned}
2\sum_{\mathrm{Re}\{\lambda_i\}>0} \lambda_i\{\bar{A} - \bar{B}\bar{D}^{-1}\bar{C}\} &= 2\ \mathrm{trace}[Y_1 N_{11}^T Y_1^{-1} + B_{11}\tilde{D}_\perp^T\tilde{D}_\perp B_{11}^T Y_1^{-1}] \\
&= \mathrm{trace}[Y_1 N_{11}^T Y_1^{-1} + N_{11} Y_1 Y_1^{-1} + 2B_{11}\tilde{D}_\perp^T\tilde{D}_\perp B_{11}^T Y_1^{-1}] \\
&= \mathrm{trace}[(Y_1 N_{11}^T + N_{11} Y_1 \\
&\qquad + B_{11}\tilde{D}_\perp^T\tilde{D}_\perp B_{11}^T) Y_1^{-1} + B_{11}\tilde{D}_\perp^T\tilde{D}_\perp B_{11}^T Y_1^{-1}] \\
&= \mathrm{trace}[-Y_1(\gamma^{-2}C_{11}^T C_{11} - C_{21}^T C_{21}) + B_{11}\tilde{D}_\perp^T\tilde{D}_\perp B_{11}^T Y_1^{-1}] \\
&= \mathrm{trace}[B_1\tilde{D}_\perp^T\tilde{D}_\perp B_1^T Y_\infty^\dagger - Y_\infty(\gamma^{-2}C_1^T C_1 - C_2^T C_2)]
\end{aligned}$$

as claimed, where we have used Lemma 3.5.7(iii)Y_1^{-1} again to obtain the fourth equality, and the fact that $Y_\infty^\dagger = \begin{bmatrix} Y_1^{-1} & 0 \\ 0 & 0 \end{bmatrix}$ to obtain the fifth. □

It is now a simple matter to piece together the intermediate results to establish the entropy formula (3.16). Taking (3.20) with Lemma 3.5.8 and (3.19), gives

$$\begin{aligned}
I(H_{ME\infty};\gamma;\infty) &= -\gamma^2\ \mathrm{trace}[\bar{D}^{-1}\bar{C}\bar{B}] - 2\gamma^2 \sum_{\mathrm{Re}\{\lambda_i\}>0} \lambda_i\{\bar{A} - \bar{B}\bar{D}^{-1}\bar{C}\} \\
&= -\gamma^2\ \mathrm{trace}[C_2 B_1 D_{21}^T + \hat{C}_2\hat{B}_1 + C_3 B_1 \tilde{D}_\perp^T + \hat{C}_3\hat{B}_1 D_{21}\tilde{D}_\perp^T] \\
&\quad - \gamma^2\ \mathrm{trace}[B_1\tilde{D}_\perp^T\tilde{D}_\perp B_1^T Y_\infty^\dagger - Y_\infty(\gamma^{-2}C_1^T C_1 - C_2^T C_2)] \\
&= -\gamma^2\ \mathrm{trace}[C_2 B_1 D_{21}^T - (C_2 + \gamma^{-2} D_{21} B_1^T X_\infty) Z_\infty (B_1 D_{21}^T + Y_\infty C_2^T)] \\
&\quad - \gamma^2\ \mathrm{trace}[-\tilde{D}_\perp B_1^T Y_\infty^\dagger B_1 \tilde{D}_\perp^T + \tilde{D}_\perp B_1^T Y_\infty^\dagger B_1 \tilde{D}_\perp^T \\
&\qquad - Y_\infty(\gamma^{-2}C_1^T C_1 - C_2^T C_2)] \\
&= \mathrm{trace}[C_1 Y_\infty C_1^T] \\
&\quad + \gamma^2\ \mathrm{trace}[(C_2 Z_\infty - C_2 + \gamma^{-2} D_{21} B_1^T X_\infty Z_\infty)(B_1 D_{21}^T + Y_\infty C_2^T)] \\
&= \mathrm{trace}[C_1 Y_\infty C_1^T] \\
&\quad + \mathrm{trace}[(D_{21} B_1^T + C_2 Y_\infty) X_\infty Z_\infty (B_1 D_{21}^T + Y_\infty C_2^T)]
\end{aligned}$$

as claimed, and the proof of the first entropy formula of Theorem 3.5.1 is complete. □

3.6 Upper Bounds and an $\mathcal{H}_\infty$/LQG Tradeoff

In Section 2.4 we saw how the entropy is related to the LQG cost. Here we shall expand on this. We will also show that the minimum value of the entropy is a monotonically decreasing function of γ, and hence obtain an important $\mathcal{H}_\infty$/LQG tradeoff. To begin with, we shall for convenience collect together the upper bounds implied by the Minimum Entropy $\mathcal{H}_\infty$ Control Problem and Theorem 2.4.4.

Proposition 3.6.1 (Upper bounds on $\mathcal{H}_\infty$-norm and LQG cost) *Suppose $K_{ME\infty}$ solves the Minimum Entropy $\mathcal{H}_\infty$ Control Problem (Problem 3.2.6), in the case of entropy evaluated at infinity, for a given standard plant P and $\gamma > \gamma_o$. Let $H_{ME\infty} = \mathcal{F}(P, K_{ME\infty})$ be the corresponding minimum entropy closed-loop. Then the $\mathcal{H}_\infty$-norm of $H_{ME\infty}$ and the LQG cost of $H_{ME\infty}$ satisfy*

(i) $\|H_{ME\infty}\|_\infty < \gamma$.

(ii) $C(H_{ME\infty}) \leq I(H_{ME\infty}; \gamma; \infty)$.

(iii) $I(H_{ME\infty}; \gamma; \infty) - C(H_{ME\infty}) = O(\gamma^{-2})$.

Equality is achieved in (ii) when $\gamma \to \infty$.

As described in Section 2.5, the bound on the $\mathcal{H}_\infty$-norm gives a robustness guarantee via the Small Gain Theorem; the robustness guarantee decreases with increasing γ. On the other hand, the other bound is an upper bound on the LQG cost. The next theorem tells us that these upper bounds trade off against each other: robustness can only be improved at the expense of the (upper bound on the) LQG cost.

Theorem 3.6.2 (The $\mathcal{H}_\infty$/LQG tradeoff) *The minimum value of the closed-loop entropy, $I(H_{ME\infty}; \gamma; \infty)$, is a monotonically decreasing function of γ.*

Proof Suppose $\gamma_2 \geq \gamma_1 > \gamma_o$. If K_1 is a (P, γ_1)-admissible controller, then K_1 stabilizes P and $\|\mathcal{F}(P, K_1)\|_\infty < \gamma_1 \leq \gamma_2$, so K_1 is also (P, γ_2)-admissible. So the set of (P, γ_1)-admissible closed-loops is a subset of the set of (P, γ_2)-admissible closed-loops. In particular, if H_i, $i = 1, 2$, minimizes $I(H; \gamma_i; \infty)$ over all (P, γ_i)-admissible closed-loops H, then H_1 is a (P, γ_1)-admissible closed-loop and a (P, γ_2)-admissible closed-loop. Therefore

$$I(H_1; \gamma_2; \infty) \geq I(H_2; \gamma_2; \infty).$$

But from Proposition 2.3.2,

$$I(H_1; \gamma_1; \infty) \geq I(H_1; \gamma_2; \infty),$$

so

$$I(H_1; \gamma_1; \infty) \geq I(H_2; \gamma_2; \infty),$$

and the result is proved. □

3.7 Recovery of the LQG Solution

The previous section showed us that loosening the $\mathcal{H}_\infty$-norm bound (by increasing γ), brings about a monotonic decrease in the entropy. By Theorem 2.4.4, if the $\mathcal{H}_\infty$-norm bound is completely relaxed ($\gamma \to \infty$) then the entropy becomes the LQG cost, and the set of (P, γ)-admissible controllers becomes the set of all controllers which stabilize P. We have therefore established the following.

Theorem 3.7.1 *The following two problems are equivalent:*

(i) *The Minimum Entropy $\mathcal{H}_\infty$ Control Problem (Problem 3.2.6), in the case of $\gamma \to \infty$ and entropy evaluated at infinity.*

(ii) *The LQG Control Problem (Problem 3.2.8).*

In allowing $\gamma \to \infty$, any robustness guaranteed by the Minimum Entropy $\mathcal{H}_\infty$ Control Problem is forfeited to obtain LQG-optimality. As pointed out by [15], an LQG-optimal controller has no guaranteed robustness margins. But if we are prepared to sacrifice a little LQG performance, some robustness can be guaranteed by using a minimum entropy $\mathcal{H}_\infty$ controller. Of course, it is in any case of interest to know the LQG-optimal (*resp.* $\mathcal{H}_\infty$-optimal) solution, for this sets the achievable limit of LQG performance (*resp.* robustness).

Suppose now that $D_{11} = 0$, so that the minimum value of the entropy and the minimum value of the LQG cost are finite. Theorem 3.7.1 allows us to recover the state-space solution to the LQG problem as the limit as $\gamma \to \infty$ of the state-space solution of the Minimum Entropy $\mathcal{H}_\infty$ Control Problem given in Remark 3.5.6 and Theorem 3.5.1. Although the solution to the LQG problem is well-known (see e.g., [2, 38]), it is worth checking that we do obtain it in the limit.

If X_∞ and Y_∞ are non-singular then we can apply the main result of [62] to the stabilizing solutions X_∞^{-1} and Y_∞^{-1} of the algebraic Riccati equations $X_\infty^{-1}(3.14)X_\infty^{-1}$ and $Y_\infty^{-1}(3.15)Y_\infty^{-1}$. This shows that both X_∞ and Y_∞ are monotonically decreasing with γ: for $\gamma_2 \geq \gamma_1 > \gamma_o$ and with an obvious notation

$$X_\infty(\gamma_2) \leq X_\infty(\gamma_1) \qquad \text{and} \qquad Y_\infty(\gamma_2) \leq Y_\infty(\gamma_1).$$

(If X_∞ and Y_∞ are not both non-singular then the above argument can be applied after identifying the non-singular parts of X_∞ and Y_∞ in the manner described in [29]. The same monotonic behaviour is obtained.)

Allowing $\gamma \to \infty$ in equations (3.14) and (3.15) gives

$$X_\infty(\infty) = X_2 \qquad \text{and} \qquad Y_\infty(\infty) = Y_2,$$

where $X_2 = X_2^T \geq 0$ is the solution to the LQG algebraic Riccati equation

$$0 = X_2(A - B_2D_{12}^TC_1) + (A - B_2D_{12}^TC_1)^TX_2 + C_1^TD_\perp D_\perp^TC_1 - X_2B_2B_2^TX_2$$

such that

$$A - B_2D_{12}^TC_1 - B_2B_2^TX_2 \quad \text{is asymptotically stable,}$$

and $Y_2 = Y_2^T \geq 0$ is the solution to the LQG algebraic Riccati equation

$$0 = Y_2(A - B_1D_{21}^TC_2)^T + (A - B_1D_{21}^TC_2)Y_2 + B_1\tilde{D}_\perp^T\tilde{D}_\perp B_1^T - Y_2C_2^TC_2Y_2$$

such that

$$A - B_1D_{21}^TC_2 - Y_2C_2^TC_2 \quad \text{is asymptotically stable.}$$

The final step is to note that $\lim_{\gamma\to\infty}\{Z_\infty(\gamma)\} = I$: then taking the limit in Remark 3.5.6 we obtain the LQG controller

$$K_{LQG} = \left[\begin{array}{c|c} A - B_1D_{21}^TC_2 - B_2D_{12}^TC_1 - Y_2C_2^TC_2 - B_2B_2^TX_2 & B_1D_{21}^T + Y_2C_2^T \\ \hline -D_{12}^TC_1 - B_2^TX_2 & 0 \end{array}\right],$$

and taking the limit in Theorem 3.5.1 we obtain the minimum LQG cost

$$C(H_{LQG}) = \text{trace}[B_1^TX_2B_1] + \text{trace}[(D_{12}^TC_1 + B_2^TX_2)Y_2(C_1^TD_{12} + X_2B_2)].$$

We shall need these state-space formulae later.

Chapter 4

The Minimum Entropy $\mathcal{H}_\infty$ Distance Problem

4.1 Introduction

In the previous chapter we solved the Minimum Entropy $\mathcal{H}_\infty$ Control Problem directly in terms of the given standard plant. Here we will provide an alternative solution via an associated Minimum Entropy $\mathcal{H}_\infty$ Distance Problem. The solution to this problem is interesting in its own right, not least because it gives us an appreciation of the underlying structure of the control problem. Indeed, it is only within the last year or two that it has been possible to solve $\mathcal{H}_\infty$ control problems without solving an associated $\mathcal{H}_\infty$ General Distance Problem. See [12] for details of the role of the General Distance Problem in $\mathcal{H}_\infty$-optimal control.

Our solution of the Minimum Entropy $\mathcal{H}_\infty$ Distance Problem uses the parametrization of all error systems satisfying the $\mathcal{L}_\infty$-norm bound. The results are reminiscent of the solution of the Minimum Entropy $\mathcal{H}_\infty$ Control Problem, although there are some extra subtleties in the proofs. We find that the central member of the admissible class minimizes the entropy at infinity and an explicit state-space formula is derived for the minimum value of the entropy at infinity (when it is finite). A link with abstract 'band extension' problems is given.

4.2 From $\mathcal{H}_\infty$ Control Problem to $\mathcal{H}_\infty$ Distance Problem

The transformation from an $\mathcal{H}_\infty$ control problem to its equivalent $\mathcal{H}_\infty$ General Distance Problem is by now well-known. The details may be found in [12]. We follow the same steps on our Minimum Entropy $\mathcal{H}_\infty$ Control Problem to reduce it to an equivalent Minimum Entropy $\mathcal{H}_\infty$ Distance Problem.

We take the initial problem setup to be exactly as in Section 3.2, Figure 3.1, and Problem 3.2.6. That is, we are given $s_0 \in (0, \infty]$, $\gamma > \gamma_o$ and a standard plant

$$P = \begin{matrix} & \overset{m_1}{\longleftrightarrow} & \overset{m_2}{\longleftrightarrow} \\ p_1 \updownarrow & P_{11} & P_{12} \\ p_2 \updownarrow & P_{21} & P_{22} \end{matrix} \quad ; \qquad m_1 \geq p_2, \; p_1 \geq m_2.$$

The aim is to minimize the closed-loop entropy, $I(\mathcal{F}(P,K);\gamma;s_0)$, over all stabilizing controllers K which satisfy

$$\|\mathcal{F}(P,K)\|_\infty < \gamma. \tag{4.1}$$

Use the parametrization of all stabilizing controllers of [64, 37] to reduce (4.1) to the equivalent *model-matching problem* of finding $\hat{Q} \in \mathcal{RH}_\infty$ such that

$$\|T_1 + T_2\hat{Q}T_3\|_\infty < \gamma, \tag{4.2}$$

and then exploit the unitary invariance of the $\mathcal{L}_\infty$-norm to reduce (4.2) to the $\mathcal{H}_\infty$ *General Distance Problem* of finding $\hat{Q} \in \mathcal{RH}_\infty$ such that

$$\left\| \begin{bmatrix} R_{11} & R_{12} \\ R_{21} & R_{22} + \hat{Q} \end{bmatrix} \right\|_\infty < \gamma,$$

where

$$R = \begin{matrix} & \overset{m_1-p_2}{\longleftrightarrow} & \overset{m_2}{\longleftrightarrow} \\ \begin{matrix} p_1-m_2 \updownarrow \\ p_2 \updownarrow \end{matrix} & \left[\begin{matrix} R_{11} \\ R_{21} \end{matrix}\right. & \left.\begin{matrix} R_{12} \\ R_{22} \end{matrix}\right] \end{matrix}$$

is anticausal (i.e., $R^* \in \mathcal{RH}_\infty$) and is known in terms of the standard plant P (see e.g., [16, Theorem 3.2.6]).

Define the *error system* E by

$$E := \begin{bmatrix} R_{11} & R_{12} \\ R_{21} & R_{22} + \hat{Q} \end{bmatrix}. \tag{4.3}$$

Then $I(\mathcal{F}(P, K); \gamma; s_0) = I(E; \gamma; s_0)$ because entropy is unitarily invariant (from Proposition 2.3.1(vi)). Hence, the closed-loop transfer function $\mathcal{F}(P, K)$ and the error system E have the same entropy, as well as the same $\mathcal{L}_\infty$-norm. This allows us to solve our original Minimum Entropy $\mathcal{H}_\infty$ Control Problem by solving the following equivalent error system Minimum Entropy $\mathcal{H}_\infty$ Distance Problem.

Problem 4.2.1 (The Minimum Entropy $\mathcal{H}_\infty$ Distance Problem)
Let $s_0 \in (0, \infty]$ and let

$$R = \begin{matrix} & \overset{m_1-p_2}{\longleftrightarrow} & \overset{m_2}{\longleftrightarrow} \\ \begin{matrix} p_1-m_2 \updownarrow \\ p_2 \updownarrow \end{matrix} & \left[\begin{matrix} R_{11} \\ R_{21} \end{matrix}\right. & \left.\begin{matrix} R_{12} \\ R_{22} \end{matrix}\right] \end{matrix}$$

be given, where

$$R^* \in \mathcal{RH}_\infty, \quad m_1 \geq p_2, \quad p_1 \geq m_2.$$

For $\hat{Q} \in \mathcal{RH}_\infty$, define the error system E by

$$E := \begin{bmatrix} R_{11} & R_{12} \\ R_{21} & R_{22} + \hat{Q} \end{bmatrix},$$

and let

$$\gamma_o = \inf_{\hat{Q}} \{ \|E\|_\infty : \quad \hat{Q} \in \mathcal{RH}_\infty \}.$$

Then for $\gamma > \gamma_o$, find $\hat{Q} \in \mathcal{RH}_\infty$ such that the entropy $I(E; \gamma; s_0)$ is minimized over those E which satisfy $\|E\|_\infty < \gamma$.

Of course, by construction γ_o above is exactly the γ_o defined in the $\mathcal{H}_\infty$-optimal control problem (Problem 3.2.5).

4.3 Relations to the Band Extension Problem

The Minimum Entropy $\mathcal{H}_\infty$ Distance Problem is related to signal processing via the problem of finding the positive definite 'band extension' of a given operator. To see the connection, use a well-known fact [32, Theorem 7.7.6] and the definition of the $\mathcal{L}_\infty$-norm to show that

$$\left\| \begin{bmatrix} R_{11} & R_{12} \\ R_{21} & R_{22}+\hat{Q} \end{bmatrix} \right\|_\infty < \gamma$$

if and only if

$$M := \begin{bmatrix} I & 0 & \gamma^{-1}R_{21} & \gamma^{-1}(R_{22}+\hat{Q}) \\ 0 & I & \gamma^{-1}R_{11} & \gamma^{-1}R_{12} \\ \gamma^{-1}R_{21}^* & \gamma^{-1}R_{11}^* & I & 0 \\ \gamma^{-1}(\hat{Q}+R_{22})^* & \gamma^{-1}R_{12}^* & 0 & I \end{bmatrix} > 0, \quad \forall s = j\omega. \tag{4.4}$$

Thus we seek a positive definite extension M of the 'band' data

$$N := \begin{bmatrix} I & 0 & \gamma^{-1}R_{21} & \gamma^{-1}R_{22} \\ 0 & I & \gamma^{-1}R_{11} & \gamma^{-1}R_{12} \\ \gamma^{-1}R_{21}^* & \gamma^{-1}R_{11}^* & I & 0 \\ \gamma^{-1}R_{22}^* & \gamma^{-1}R_{12}^* & 0 & I \end{bmatrix}.$$

This can be interpreted as 'band' data because only the anticausal component of R_{22} is specified. In [21, 22] the *band extension* of N is defined as $M > 0$ in (4.4) such that M^{-1} has the same banded structure as N. It is shown in [21, 22] that this unique band extension also minimizes the entropy. Although the very general results of [21, 22] could be applied to our problem, we choose to adapt the method of [4, 3], as done in the previous chapter. This makes for a relatively short and self-contained derivation.

4.4 Solution in the General Case

In this section we solve the Minimum Entropy $\mathcal{H}_\infty$ Distance Problem at an arbitrary $s_0 \in (0,\infty)$. Solution proceeds by firstly parametrizing all error systems E which satisfy the bound $\|E\|_\infty < \gamma$. Such a parametrization is given in [5], in terms of a linear fractional map of a *J-unitary* matrix and an arbitrary stable contraction Φ, but it is more convenient to use the parametrization of [30] in terms of the linear fractional map of an *all-pass* matrix and an arbitrary stable contraction Φ.

To begin, we parametrize the class of error systems E over which the entropy must be minimized.

Lemma 4.4.1 ([30]) *All solutions*

$$E = \begin{matrix} & \overset{m_1-p_2}{\longleftrightarrow} & \overset{p_2}{\longleftrightarrow} \\ {\scriptstyle p_1-m_2}\updownarrow & R_{11} & R_{12} \\ {\scriptstyle m_2}\updownarrow & R_{21} & R_{22}+\hat{Q} \end{matrix}$$

with

$$R^* \in \mathcal{RH}_\infty \text{ and } \hat{Q} \in \mathcal{RH}_\infty, \qquad m_1 \geq p_2 \text{ and } p_1 \geq m_2,$$

to the Distance Problem $\|E\|_\infty < \gamma$*, where* $\gamma > \gamma_o$*, are given by:*

$$E = \gamma \mathcal{F}(R_{aa} + Q_{aa}, \Psi),$$

where

$$\Psi := \begin{matrix} & \overset{m_1-p_2}{\longleftrightarrow} & \overset{p_2}{\longleftrightarrow} \\ {\scriptstyle p_1-m_2}\updownarrow & 0 & 0 \\ {\scriptstyle m_2}\updownarrow & 0 & \Phi \end{matrix}, \quad \Phi \in \mathcal{BRH}_\infty. \tag{4.5}$$

Also,

$$R_{aa} + Q_{aa} := \begin{matrix} & \overset{m_1}{\longleftrightarrow} & \overset{p_1}{\longleftrightarrow} \\ {\scriptstyle p_1}\updownarrow & [R_{aa}+Q_{aa}]_{11} & [R_{aa}+Q_{aa}]_{12} \\ {\scriptstyle m_1}\updownarrow & [R_{aa}+Q_{aa}]_{21} & [R_{aa}+Q_{aa}]_{22} \end{matrix}$$

$$= \begin{matrix} & \overset{m_1-p_2}{\longleftrightarrow} & \overset{p_2}{\longleftrightarrow} & \overset{p_1-m_2}{\longleftrightarrow} & \overset{m_2}{\longleftrightarrow} \\ {\scriptstyle p_1-m_2}\updownarrow & \gamma^{-1}R_{11} & \gamma^{-1}R_{12} & R_{13} & 0 \\ {\scriptstyle m_2}\updownarrow & \gamma^{-1}R_{21} & \gamma^{-1}(R_{22}+Q_{22}) & R_{23}+Q_{23} & Q_{24} \\ {\scriptstyle m_1-p_2}\updownarrow & R_{31} & R_{32}+Q_{32} & R_{33}+Q_{33} & Q_{34} \\ {\scriptstyle p_2}\updownarrow & 0 & Q_{42} & Q_{43} & Q_{44} \end{matrix}.$$

Furthermore, $R_{ij}^* \in \mathcal{RH}_\infty$*,* $Q_{ij} \in \mathcal{RH}_\infty$*,* $R_{aa} + Q_{aa}$ *is all-pass and* $Q_{44}(\infty) = 0$.

State-space realizations of R_{ij} and Q_{ij} are available in [30], in terms of the realization of R and the solutions to two algebraic Riccati equations. These realizations will be stated and used in Section 4.5

The next lemma relates the entropy of the linear fractional map of an all-pass matrix V and an arbitrary stable contraction Ψ to the entropy of Ψ itself.

Lemma 4.4.2 *Suppose*

$$V = \begin{bmatrix} V_{11} & V_{12} \\ V_{21} & V_{22} \end{bmatrix} \in \mathcal{RL}_\infty$$

is all-pass, $\Psi \in \mathcal{BRH}_\infty$ *and* $V_{22}\Psi \in \mathcal{RH}_\infty$*. Then*

$$\begin{aligned} I(\gamma\mathcal{F}(V,\Psi);\gamma;s_0) &= \gamma^2 I(\Psi;1;s_0) + \gamma^2 I(V_{11};1;s_0) \\ &\quad + \gamma^2 s_0 \ln|\det(I - V_{22}(s_0)\Psi(s_0))|. \end{aligned}$$

Proof Since V_{22} is part of an all-pass matrix, $\|V_{22}\|_\infty \leq 1$. Therefore $\|V_{22}\Psi\|_\infty < 1$ because $\|\Psi\|_\infty < 1$. This, together with the assumption that $V_{22}\Psi \in \mathcal{RH}_\infty$, implies that $\det(I - V_{22}\Psi)^{\pm 1} \in \mathcal{RH}_\infty$. As V is all-pass, we then have that [51]

$$[\gamma\mathcal{F}(V,\Psi)]^*[\gamma\mathcal{F}(V,\Psi)] < \gamma^2 I,$$

so the entropy $I(\gamma\mathcal{F}(V,\Psi);\gamma;s_0)$ is well-defined. Using

$$V^*V = I \tag{4.6}$$

in block-partitioned form, it is straightforward to show that

$$I - \gamma^{-2}[\gamma\mathcal{F}(V,\Psi)]^*[\gamma\mathcal{F}(V,\Psi)] = V_{21}^*[I - V_{22}\Psi]^{-*}[I - \Psi^*\Psi][I - V_{22}\Psi]^{-1}V_{21}.$$

From this, and the fact that for any square real-rational transfer function matrix G

$$\ln|\det(G^*G)| = \ln|\det G| + \ln|\det G^*| = 2\ln|\det G|, \quad \forall s = j\omega$$

we obtain, $\forall s = j\omega$

$$\begin{aligned}\ln|\det(I - \gamma^{-2}[\gamma\mathcal{F}(V,\Psi)]^*[\gamma\mathcal{F}(V,\Psi)])| &= \ln|\det(I - \Psi^*\Psi)| \\ &\quad + \ln|\det(V_{21}^*V_{21})| - 2\ln|\det(I - V_{22}\Psi)|.\end{aligned}$$

Substituting this into the definition of $I(\gamma\mathcal{F}(V,\Psi);\gamma;s_0)$ and using the (1,1) block of equation (4.6) to write $V_{21}^*V_{21} = I - V_{11}^*V_{11}$ it follows that

$$\begin{aligned}I(\gamma\mathcal{F}(V,\Psi);\gamma;s_0) &= \gamma^2 I(\Psi;1;s_0) + \gamma^2 I(V_{11};1;s_0) \\ &\quad + \frac{\gamma^2}{\pi}\int_{-\infty}^{\infty} \ln|\det(I - V_{22}(j\omega)\Psi(j\omega))|\left[\frac{s_0^2}{s_0^2+\omega^2}\right]d\omega.\end{aligned} \tag{4.7}$$

But, from above, $\det(I - V_{22}\Psi)^{\pm 1} \in \mathcal{RH}_\infty$, which permits the use of Poisson's Integral Theorem [52, Theorem 17.16] to evaluate the integral in (4.7), giving

$$\begin{aligned}I(\gamma\mathcal{F}(V,\Psi);\gamma;s_0) &= \gamma^2 I(\Psi;1;s_0) + \gamma^2 I(V_{11};1;s_0) \\ &\quad + \gamma^2 s_0 \ln|\det(I - V_{22}(s_0)\Psi(s_0))|\end{aligned}$$

as claimed. □

We are now in a position to derive the unique stable contractive Φ in the parametrization of all error systems, which minimizes the entropy $I(E;\gamma;s_0)$.

Theorem 4.4.3 *Consider the class of error systems E which satisfy the condition $\|E\|_\infty < \gamma$ as parametrized in Lemma 4.4.1 by*

$$E = \gamma\mathcal{F}(R_{aa} + Q_{aa}, \begin{bmatrix} 0 & 0 \\ 0 & \Phi \end{bmatrix}), \quad \Phi \in \mathcal{BRH}_\infty. \tag{4.8}$$

Then the entropy $I(E;\gamma;s_0)$ is minimized over this class of E by the unique choice

$$\Phi = Q_{44}^*(s_0).$$

Proof Here we adapt the approach of [4, 3] to the present setting. Lemma 4.4.1 gives all error systems in the form (4.8), where $R_{aa} + Q_{aa}$ is all-pass. Also,

$$[R_{aa} + Q_{aa}]_{22} \begin{bmatrix} 0 & 0 \\ 0 & \Phi \end{bmatrix} = \begin{bmatrix} 0 & Q_{34}\Phi \\ 0 & Q_{44}\Phi \end{bmatrix} \in \mathcal{RH}_\infty,$$

because Q_{34}, Q_{44} and Φ are all in $\mathcal{RH}_\infty$. Hence, we may apply Lemma 4.4.2 to E to obtain

$$\begin{aligned} I(E;\gamma;s_0) &= \gamma^2 I(\Phi;1;s_0) + \gamma^2 I([R_{aa}+Q_{aa}]_{11};1;s_0) \\ &\quad + \gamma^2 s_0 \ln|\det(I - Q_{44}(s_0)\Phi(s_0))|. \end{aligned} \tag{4.9}$$

Define a constant Julia matrix [65, p148] by

$$U = \begin{bmatrix} U_{11} & U_{12} \\ U_{21} & U_{22} \end{bmatrix} := \begin{bmatrix} -Q_{44}^*(s_0) & (I - Q_{44}^*(s_0)Q_{44}(s_0))^{1/2} \\ (I - Q_{44}(s_0)Q_{44}^*(s_0))^{1/2} & Q_{44}(s_0) \end{bmatrix}.$$

It is easy to verify that U is unitary. Also,

$$U_{22}\Phi(s) = Q_{44}(s_0)\Phi(s)$$

which is in $\mathcal{RH}_\infty$. Now map the unit ball in $\mathcal{RH}_\infty$ onto itself by the linear fractional map

$$\tilde{\Phi} := \mathcal{F}(U, \Phi), \quad \Phi \in \mathcal{BRH}_\infty.$$

Note that, under this mapping,

$$\Phi = Q_{44}^*(s_0) \quad \Longleftrightarrow \quad \tilde{\Phi} = 0. \tag{4.10}$$

Lemma 4.4.2 is applicable:

$$I(\tilde{\Phi};1;s_0) = I(\Phi;1;s_0) + I(Q_{44}^*(s_0);1;s_0) + s_0 \ln|\det(I - Q_{44}(s_0)\Phi(s_0))|.$$

Use this together with (4.9) to relate the entropy of E to the entropy of $\tilde{\Phi}$:

$$I(E;\gamma;s_0) = \gamma^2 I(\tilde{\Phi};1;s_0) + \gamma^2 I([R_{aa}+Q_{aa}]_{11};1;s_0) - \gamma^2 I(Q_{44}^*(s_0);1;s_0), \tag{4.11}$$

from which it is immediate that $I(E;\gamma;s_0)$ is minimized by the unique choice $\tilde{\Phi} = 0$. But from (4.10), $\tilde{\Phi} = 0 \iff \Phi = Q_{44}^*(s_0)$, and the theorem is proved. □

An expression for the minimum value of the entropy now follows.

Corollary 4.4.4 *The minimum value of $I(E;\gamma;s_0)$, over the class of all error systems E satisfying $\|E\|_\infty < \gamma$, is given by*

$$\begin{aligned} I(E_{MEs_0};\gamma;s_0) &= \gamma^2 I([R_{aa}+Q_{aa}]_{11};1;s_0) - \gamma^2 I(Q_{44}^*(s_0);1;s_0) && (4.12) \\ &= \gamma^2 s_0 \{-\ln|\det R_{31}^*(s_0)| - \ln|\det Q_{42}(s_0)| \\ &\quad + (1/2)\ln|\det(I - Q_{44}^*(s_0)Q_{44}(s_0))|\}. && (4.13) \end{aligned}$$

Proof Equation (4.12) follows immediately from equation (4.11) on setting $\tilde{\Phi} = 0$. To show (4.13), recall that $R_{aa} + Q_{aa}$ is all-pass i.e.,

$$(R_{aa} + Q_{aa})^*(R_{aa} + Q_{aa}) = I.$$

The (1,1) block of this gives,

$$I - [R_{aa} + Q_{aa}]_{11}^*[R_{aa} + Q_{aa}]_{11} = [R_{aa} + Q_{aa}]_{21}^*[R_{aa} + Q_{aa}]_{21}$$

so that, along the imaginary axis,

$$\begin{aligned} \ln|\det(I - [R_{aa} + Q_{aa}]_{11}^*[R_{aa} + Q_{aa}]_{11})| &= 2\ln|\det([R_{aa} + Q_{aa}]_{21})| \quad (4.14) \\ &= 2\ln|\det R_{31}^*| + 2\ln|\det Q_{42}|, \quad (4.15) \end{aligned}$$

where (4.15) follows from (4.14) on examination of the structure of $R_{aa} + Q_{aa}$ in Lemma 4.4.1. Substituting (4.15) into the definition of entropy, we see that

$$\begin{aligned} \gamma^2 I([R_{aa} + Q_{aa}]_{11}; 1; s_0) &= -\frac{\gamma^2}{\pi}\int_{-\infty}^{\infty} \{\ln|\det R_{31}^*(j\omega)| \\ &\quad + \ln|\det Q_{42}(j\omega)|\}\left[\frac{s_0^2}{s_0^2 + \omega^2}\right] d\omega. \quad (4.16) \end{aligned}$$

Since $R_{31}^{\pm *} \in \mathcal{RH}_\infty$ and $Q_{42}^{\pm 1} \in \mathcal{RH}_\infty$ [30], Poisson's Integral Theorem may be used to evaluate (4.16) as

$$\gamma^2 I([R_{aa} + Q_{aa}]_{11}; 1; s_0) = -\gamma^2 s_0 (\ln|\det R_{31}^*(s_0)| + \ln|\det Q_{42}(s_0)|). \quad (4.17)$$

The second term in (4.12) is

$$\begin{aligned} -\gamma^2 I(Q_{44}^*(s_0); 1; s_0) &= \frac{\gamma^2}{2\pi}\ln|\det(I - Q_{44}(s_0)Q_{44}^*(s_0))| \int_{-\infty}^{\infty}\left[\frac{s_0^2}{s_0^2 + \omega^2}\right] d\omega \\ &= \frac{\gamma^2}{2\pi}\ln|\det(I - Q_{44}(s_0)Q_{44}^*(s_0))|(\pi s_0) \end{aligned}$$

and this with (4.17) gives (4.13). □

4.5 Solution at Infinity

We turn our attention in this section to the special case of entropy at infinity, i.e., when $s_0 \to \infty$. The results for the minimum entropy problem at infinity are particularly simple, and are similar to those derived in Chapter 3 for the Minimum Entropy $\mathcal{H}_\infty$ Control Problem. The minimum entropy solution is just the central solution, and an explicit formula for the minimum value of the entropy is derived in terms of the state-space realizations inherent in the solution of the Distance Problem of Lemma 4.4.1; these state-space realizations are stated in the next lemma.

The state-space formulae assume that $R(\infty) = 0$. This is a sufficient condition for $I(E_{ME\infty}; \gamma; \infty)$ to be finite. $R(\infty) = 0$ is implied by the assumption that $P_{11}(\infty) = 0$, made in Section 3.4 to ensure that $I(H_{ME\infty}; \gamma; \infty)$ is finite. See Remark 3.4.1.

Lemma 4.5.1 **([30])** *Suppose R has a realization*

$$R = \begin{matrix} \\ 2n \updownarrow \\ p_1-m_2 \updownarrow \\ m_2 \updownarrow \end{matrix} \begin{matrix} \overset{2n}{\longleftrightarrow} & \overset{m_1-p_2}{\longleftrightarrow} & \overset{p_2}{\longleftrightarrow} \\ \end{matrix} \left[\begin{array}{c|cc} A & B_1 & B_2 \\ \hline C_1 & 0 & 0 \\ C_2 & 0 & 0 \end{array}\right],$$

where R is anticausal i.e., $-A$ is asymptotically stable. Then $R_{aa} + Q_{aa}$, as in the parametrization of all solutions to the Distance Problem $\|E\|_\infty < \gamma$ given in Lemma 4.4.1, has a realization

$$R_{aa} + Q_{aa} = \left[\begin{array}{c|c} \tilde{A} & \tilde{B} \\ \hline \tilde{C} & \tilde{D} \end{array}\right]$$

where

$$\tilde{A} := \begin{bmatrix} A & 0 \\ 0 & \hat{A} \end{bmatrix}$$

$$\tilde{B} := \begin{bmatrix} \gamma^{-1/2}B_1 & \gamma^{-1/2}B_2 & -\gamma^{-3/2}XC_1^T & 0 \\ 0 & \gamma^{-3/2}YZ^{-1}B_2 & -\gamma^{-1/2}C_1^T & \gamma^{-1/2}Z^{-T}C_2^T \end{bmatrix}$$

$$\tilde{C} := \begin{bmatrix} \gamma^{-1/2}C_1 & 0 \\ \gamma^{-1/2}C_2 & -\gamma^{-3/2}C_2X \\ -\gamma^{-3/2}B_1^TY & \gamma^{-1/2}B_1^TZ^T \\ 0 & -\gamma^{-1/2}B_2^T \end{bmatrix}$$

$$\tilde{D} := \begin{bmatrix} 0 & 0 & I & 0 \\ 0 & 0 & 0 & I \\ I & 0 & 0 & 0 \\ 0 & I & 0 & 0 \end{bmatrix}$$

and where $X = X^T$ solves

$$0 = XA^T + AX + \gamma^{-2}XC_1^TC_1X + B_1B_1^T + B_2B_2^T \tag{4.18}$$

such that

$$-(A + \gamma^{-2}XC_1^TC_1) \qquad \textit{is asymptotically stable,}$$

where $Y = Y^T$ solves

$$0 = YA + A^TY + \gamma^{-2}YB_1B_1^TY + C_1^TC_1 + C_2^TC_2 \tag{4.19}$$

such that

$$-(A + \gamma^{-2}B_1B_1^TY) \qquad \textit{is asymptotically stable,}$$

and where

$$Z := \gamma^{-2}XY - I,$$
$$\hat{A} := -A^T - \gamma^{-2}YZ^{-1}B_2B_2^T - \gamma^{-2}C_1^TC_1X.$$

It is convenient to introduce the notation

$$E\{\Phi\} := \gamma\mathcal{F}(R_{aa} + Q_{aa}, \begin{bmatrix} 0 & 0 \\ 0 & \Phi \end{bmatrix})$$

to denote an error system produced by Lemma 4.4.1 by a $\Phi \in \mathcal{BRH}_\infty$. Also define

$$\begin{aligned} \Phi_{ME\infty} &:= \lim_{s_0 \to \infty} \{\Phi_{MEs_0}\} \\ &= \lim_{s_0 \to \infty} \{Q_{44}^*(s_0)\} \\ &= 0, \end{aligned}$$

where the second equality is from Theorem 4.4.3 and the third equality is from Lemma 4.4.1. Then

$$E\{\Phi_{ME\infty}\} = E\{0\} = \gamma[R_{aa} + Q_{aa}]_{11}. \tag{4.20}$$

We can now apply the results of the previous section, using the state-space realization given in Lemma 4.5.1, to obtain the following important theorem.

Theorem 4.5.2

(i) *The choice*

$$\Phi = \Phi_{ME\infty} = 0$$

in the parametrization of Lemma 4.4.1 of all error systems E satisfying $\|E\|_\infty < \gamma$ is the unique choice of Φ which satisfies:

$$\begin{aligned} &\forall \Psi \in \mathcal{BRH}_\infty \quad (\Psi \neq 0): \\ &\exists s(\Psi) < \infty \quad \textit{such that} \quad \forall s_0 > s(\Psi): \quad I(E\{\Phi_{ME\infty}\}; \gamma; s_0) < I(E\{\Psi\}; \gamma; s_0). \end{aligned}$$

Thus the minimum entropy error system is $E_{ME\infty} = E\{\Phi_{ME\infty}\} = E\{0\}$.

(ii) *If $R(\infty) = 0$ then $\Phi = \Phi_{ME\infty} = 0$ is the unique choice of Φ which minimizes $I(E\{\Phi\}; \gamma; \infty)$ over $\Phi \in \mathcal{BRH}_\infty$.*

(iii) *If $R(\infty) = 0$, then the minimum entropy error system has a state-space realization*

$$\begin{aligned} E_{ME\infty} &= \gamma[R_{aa} + Q_{aa}]_{11} \\ &= \left[\begin{array}{cc|cc} A & 0 & B_1 & B_2 \\ 0 & \hat{A} & 0 & \gamma^{-1}YZ^{-1}B_2 \\ \hline C_1 & 0 & 0 & 0 \\ C_2 & -\gamma^{-1}C_2X & 0 & 0 \end{array}\right] \\ &= \underbrace{\left[\begin{array}{c|cc} A & B_1 & B_2 \\ \hline C_1 & 0 & 0 \\ C_2 & 0 & 0 \end{array}\right]}_{R} + \underbrace{\left[\begin{array}{c|cc} \hat{A} & 0 & \gamma^{-1}YZ^{-1}B_2 \\ \hline 0 & 0 & 0 \\ -\gamma^{-1}C_2X & 0 & 0 \end{array}\right]}_{\begin{bmatrix} 0 & 0 \\ 0 & \hat{Q}_{ME\infty} \end{bmatrix}} \end{aligned} \tag{4.21}$$

and the minimum value of the entropy is

$$I(E_{ME\infty};\gamma;\infty) = -\operatorname{trace}[B_1^T Y B_1] + \operatorname{trace}[B_2^T Y Z^{-1} B_2] \quad (4.22)$$
$$= -\operatorname{trace}[C_1 X C_1^T] + \operatorname{trace}[C_2 Z^{-1} X C_2^T]. \quad (4.23)$$

Proof *Parts (i) and (ii)* From Theorem 4.4.3, the minimum entropy error system, when entropy is evaluated at $s_0 \in (0, \infty)$, is characterized by the unique choice $\Phi = Q_{44}^*(s_0)$. Letting $s_0 \to \infty$ gives the minimum entropy at infinity choice as $\Phi = Q_{44}^*(\infty) = 0$, because Q_{44} is strictly proper from Lemma 4.4.1. When $R(\infty) \neq 0$, the minimum value of the entropy is not necessarily finite; the limiting argument then required is identical to that used in the proof of Theorem 3.4.2, and so it will not be repeated here.

Part (iii) That the minimum entropy error system has a realization (4.21) follows easily by using (4.20) in Lemma 4.5.1.

To obtain the minimum value of the entropy, we take the limit as $s_0 \to \infty$ of the result of Corollary 4.4.4. That is,

$$I(E_{ME\infty};\gamma;\infty) = \lim_{s_0 \to \infty} \{\gamma^2 s_0 \{-\ln|\det R_{31}^*(s_0)| - \ln|\det Q_{42}(s_0)| + (1/2)\ln|\det(I - Q_{44}^*(s_0) Q_{44}(s_0))|\}\}. \quad (4.24)$$

From the proof of Lemma A.4.1 of Appendix A, we have, for a typical term from (4.24):

$$\lim_{s_0 \to \infty} \{-s_0 \ln|\det(I + \bar{C}(s_0 I - \bar{A})^{-1} \bar{B})|\} = -\operatorname{trace}[\bar{C}\bar{B}].$$

Apply this to the terms in (4.24) using

$$R_{31}^*(s_0) = I + \gamma^{-2} B_1^T (s_0 I + A^T)^{-1} Y B_1$$
$$Q_{42}(s_0) = I - \gamma^{-2} B_2^T (s_0 I - \hat{A})^{-1} Y Z^{-1} B_2$$

from Lemma 4.4.1, to get

$$I(E_{ME\infty};\gamma;\infty) = -\operatorname{trace}[B_1^T Y B_1] + \operatorname{trace}[B_2^T Y Z^{-1} B_2]$$

as claimed. Note that the third term in (4.24) is zero in the limit because Q_{44} is strictly proper.

The dual expression (4.23) follows in an entirely similar fashion; one recalls from Proposition 2.3.1(vii) that $E_{ME\infty}^T$ has the same entropy as $E_{ME\infty}$, leading to

$$I(E_{ME\infty};\gamma;\infty) = \lim_{s_0 \to \infty} \{\gamma^2 s_0 \{-\ln|\det R_{13}^*(s_0)| - \ln|\det Q_{24}(s_0)| + (1/2)\ln|\det(I - Q_{44}(s_0) Q_{44}^*(s_0))|\}\},$$

which gives (4.23) in the limit. □

Remark 4.5.3 Notice that the entropy formulae (4.22) and (4.23) depend only on the state-space realization of R and the solutions X and Y to the two algebraic Riccati equations (4.18) and (4.19) which are inherent in the solution to the Distance Problem. Calculation of the minimum value of the entropy therefore imposes negligible extra computational problems. Furthermore, the minimum entropy error system (4.21), being the linear fractional map of $\Phi = 0$, is simply γ times the p_1 by m_1 (1,1) block of $R_{aa}+Q_{aa}$ (from (4.20)), which is also available from the solution to the Distance Problem with no extra computation.

Remark 4.5.4 (Recovery of the $\mathcal{L}_2$-optimal solution) Recall, from Theorem 2.4.4 and Remark 2.4.3, that $I(E;\infty;\infty) = \|E\|_2^2$ for strictly proper E. Thus if we let $\gamma \to \infty$ in our minimum entropy solution we should obtain exactly the $\mathcal{L}_2$-optimal solution i.e., the error system with minimum $\mathcal{L}_2$-norm. We show here that this is indeed the case. By using [62], it may be shown that the positive semidefinite matrices $-X$ and $-Y$ are monotonically decreasing as γ increases. Taking $\gamma \to \infty$ we obtain

$$0 = X(\infty)A^T + AX(\infty) + B_1B_1^T + B_2B_2^T \tag{4.25}$$
$$0 = Y(\infty)A + A^TY(\infty) + C_1^TC_1 + C_2^TC_2 \tag{4.26}$$

and

$$Z(\infty) = -I,$$

(with an obvious notation). Equations (4.25) and (4.26) identify the matrices $-X(\infty)$ and $-Y(\infty)$ as the controllability and observability Gramians of $R(-s)$, respectively. (Remember that $R^* \in \mathcal{RH}_\infty$ so $-A$ is asymptotically stable.) Using this fact, a simple calculation shows that (assuming $R(\infty) = 0$ to ensure finiteness)

$$\begin{aligned} I(E_{ME\infty};\infty;\infty) &= -\ \mathrm{trace}[B_1^TY(\infty)B_1] +\ \mathrm{trace}[B_2^TY(\infty)Z(\infty)^{-1}B_2] \\ &= \mathrm{trace}\left[[B_1\ \ B_2]^T[-Y(\infty)][B_1\ \ B_2]\right] \\ &= \|R\|_2^2 \end{aligned} \tag{4.27}$$

Also, inspection of (4.21) leads to $\lim_{\gamma\to\infty}\{\hat{Q}_{ME\infty}\} = 0$, hence $\lim_{\gamma\to\infty}\{E_{ME\infty}\} = R$. It is well-known that the $\hat{Q} \in \mathcal{RH}_\infty$ which minimizes

$$\|E\|_2 = \left\| \begin{bmatrix} R_{11} & R_{12} \\ R_{21} & R_{22}+\hat{Q} \end{bmatrix} \right\|_2$$

is $\hat{Q} = Q_{\mathcal{L}_2} = 0$ (the $\mathcal{L}_2$-optimal solution) and in that case the $\mathcal{L}_2$-optimal error system is $E_{\mathcal{L}_2} = R$. We therefore have that $\lim_{\gamma\to\infty}\{\hat{Q}_{ME\infty}\} = \hat{Q}_{\mathcal{L}_2}$; that $\lim_{\gamma\to\infty}\{E_{ME\infty}\} = E_{\mathcal{L}_2} = R$; and from (4.27) that $I(E_{ME\infty};\infty;\infty) = \|E_{\mathcal{L}_2}\|_2^2$. This establishes the equivalence between the Minimum Entropy $\mathcal{H}_\infty$ Distance Problem when γ, $s_0 \to \infty$ and the $\mathcal{L}_2$-optimal Distance Problem. This corresponds exactly to the recovery of the LQG problem from the Minimum Entropy $\mathcal{H}_\infty$ Control Problem as in Section 3.7.

Remark 4.5.5 Note that if $\gamma \to \gamma_o$, then $E_{ME\infty} \to E_o$ ($\mathcal{L}_\infty$-optimal). Also (Remark 4.5.4), if $\gamma \to \infty$, then $E_{ME\infty} \to E_{\mathcal{L}_2}$ ($\mathcal{L}_2$-optimal). Thus γ can be used to move from $\mathcal{L}_\infty$-optimal to $\mathcal{L}_2$-optimal via the minimum entropy solutions for $\gamma_o < \gamma < \infty$. This, of course, corresponds exactly to the $\mathcal{H}_\infty$/LQG tradeoff of Section 3.6 for the original Minimum Entropy $\mathcal{H}_\infty$ Control Problem.

Chapter 5

Relations to Combined $\mathcal{H}_\infty$/LQG Control

5.1 Introduction

In this chapter we provide further results on the interplay between minimum entropy $\mathcal{H}_\infty$ control and LQG control. This will be done by proving the *equivalence* between the Minimum Entropy $\mathcal{H}_\infty$ Control Problem of Chapter 3 and the *Combined $\mathcal{H}_\infty$/LQG Problem* of [8, 9]. The Combined $\mathcal{H}_\infty$/LQG Problem is approached in [8, 9] by constructing upper bounds on the LQG cost using the solution of algebraic Riccati equations. We will show that the procedure of [8, 9], in the case of full-order controllers and with $\mathcal{H}_\infty$ and LQG criteria applied to the same closed-loop transfer function, is exactly the same as using minimum entropy $\mathcal{H}_\infty$ control. This gives an exact interpretation of the bounds used in [8, 9]. By exploiting the simplicity of the solution to the Minimum Entropy $\mathcal{H}_\infty$ Control Problem, we will also be able to obtain a useful simplification of the state-space solution to the Combined $\mathcal{H}_\infty$/LQG Problem given in [8].

5.2 The Combined $\mathcal{H}_\infty$/LQG Problem

We consider the same setup as for the Minimum Entropy $\mathcal{H}_\infty$ Control Problem in Section 3.2. We assume that $D_{11} = 0$. This assumption, as we saw in Section 3.4, ensures that the minimum value of the entropy and the minimum value of the LQG cost are both finite. It also ensures that the *auxiliary* cost (defined later) has a finite minimum value.

Thus we consider an n-state standard plant P with a state-space realization

$$P = \begin{bmatrix} P_{11} & P_{12} \\ P_{21} & P_{22} \end{bmatrix} = \left[\begin{array}{c|cc} A & B_1 & B_2 \\ \hline C_1 & 0 & D_{12} \\ C_2 & D_{21} & 0 \end{array}\right],$$

connected with an n-state feedback controller

$$K := \left[\begin{array}{c|c} \hat{A} & \hat{B} \\ \hline \hat{C} & 0 \end{array}\right],$$

as in Figure 3.1. It is easy to verify that the closed-loop transfer function $H = \mathcal{F}(P, K)$ has a state-space realization

$$H = \mathcal{F}(P,K) := \left[\begin{array}{c|c} \tilde{A} & \tilde{B} \\ \hline \tilde{C} & 0 \end{array}\right],$$

where

$$\tilde{A} = \begin{bmatrix} A & B_2\hat{C} \\ \hat{B}C_2 & \hat{A} \end{bmatrix}, \quad \tilde{B} = \begin{bmatrix} B_1 \\ \hat{B}D_{21} \end{bmatrix} \quad \text{and} \quad \tilde{C} = \begin{bmatrix} C_1 & D_{12}\hat{C} \end{bmatrix}.$$

In order to motivate the Combined $\mathcal{H}_\infty$/LQG Problem of [8, 9], recall the expression for the LQG cost associated with a stabilized closed-loop H, as given in Remark 2.4.3:

$$C(H) = \|H\|_2^2.$$

It is well-known that

$$C(H) = \text{trace}[\tilde{Q}\tilde{C}^T\tilde{C}], \tag{5.1}$$

where $\tilde{Q} = \tilde{Q}^T \geq 0$ is the solution to the Lyapunov equation

$$0 = \tilde{A}\tilde{Q} + \tilde{Q}\tilde{A}^T + \tilde{B}\tilde{B}^T. \tag{5.2}$$

In fact, $\tilde{Q}$ is just the closed-loop controllability Gramian. Now suppose we require the stabilized closed-loop H to satisfy $\|H\|_\infty < \gamma$ (that is, suppose H is a (P,γ)-admissible closed-loop). From [60], the Frequency Domain Inequality tells us that $\|H\|_\infty < \gamma$ if there exists a stabilizing solution $Q = Q^T \geq 0$ to the algebraic Riccati equation

$$0 = \tilde{A}Q + Q\tilde{A}^T + \gamma^{-2}Q\tilde{C}^T\tilde{C}Q + \tilde{B}\tilde{B}^T. \tag{5.3}$$

But it is easy to prove [8, 9, 60, 62] that any Q solving (5.3) overbounds the closed-loop controllability Gramian $\tilde{Q}$ i.e., $Q \geq \tilde{Q}$. In view of the evaluation of the LQG cost in (5.1), this motivates the definition of the *auxiliary* cost.

Definition 5.2.1 (Auxiliary cost [8, 9])
The auxiliary cost *associated with a (P,γ)-admissible closed-loop H is defined by*

$$J(H;\gamma) := \text{trace}[Q_s\tilde{C}^T\tilde{C}], \tag{5.4}$$

where $Q_s = Q_s^T \geq 0$ is the stabilizing solution of the algebraic Riccati equation (5.3).

It is immediate that $J(H;\gamma) \geq C(H)$ because of the previously noted bound $Q \geq \tilde{Q}$. Minimizing $J(H;\gamma)$ subject to H being (P,γ)-admissible therefore gives us a control problem which combines both $\mathcal{H}_\infty$ and LQG objectives. The similarities with the Minimum Entropy $\mathcal{H}_\infty$ Control Problem will be shown in the next section to be much more than superficial.

Problem 5.2.2 (The Combined $\mathcal{H}_\infty$/LQG Control Problem [8, 9])
Let P be a standard plant and let $\gamma > \gamma_o$. Minimize the auxiliary cost $J(H;\gamma)$ over all (P,γ)-admissible closed-loops H.

In the terminology of [8, 9], Problem 5.2.2 is a full-order (controller) problem with equalized $\mathcal{H}_\infty$ and LQG weights. Problems with reduced-order controller and/or non-equalized weights can also be treated within the framework of [8, 9]. However, the number and complexity of coupled Riccati equations needed for the solution then increases substantially—in that case the equivalence result of this chapter does not apply.

Assumptions 5.2.3 A number of mild assumptions will be made. We, of course, inherit Assumption 3.2.1 and Assumptions 3.2.7 from the Minimum Entropy $\mathcal{H}_\infty$ Control Problem. The assumption that $D_{11} = 0$ has already been mentioned as a simple way to guarantee that the minimum value of the entropy, minimum value of the auxiliary cost and minimum value of the LQG cost are all finite. Finally, in Section 5.4 we will assume that $D_{12}^T C_1 = 0$ and $B_1 D_{21}^T = 0$, for convenience. This eliminates cross-weighting terms which would otherwise cloud the argument, and is also assumed in [8, 9]. $D_{12}^T C_1 = 0$ corresponds to orthogonality of $C_1 x$, (where x is the state vector of P) and $D_{12}u$, whereas $B_1 D_{21}^T = 0$ corresponds to orthogonality of plant disturbance $B_1 w$ and sensor noise $D_{21}w$.

It is the purpose of this chapter to establish that under these conditions the Combined $\mathcal{H}_\infty$/LQG Problem and the Minimum Entropy $\mathcal{H}_\infty$ Control Problem are in fact *identical.* We will prove this in Section 5.3 by establishing the key result equating entropy and auxiliary cost:

$$I(H;\gamma;\infty) = J(H;\gamma).$$

This link is interesting because it draws together two apparently unconnected frameworks in a fairly deep way. In Section 5.4 state-space solutions to the two problems are given, using the results of Chapter 3 and [8], and it is shown explicitly that these solutions are identical. The minimum value of the entropy, by the results of Chapter 3, can be written in terms of the two Riccati equations needed to find the minimum entropy $\mathcal{H}_\infty$ controller. But the minimum value of the auxiliary cost, given by [8], requires a third Riccati equation which is *coupled* to the other two. Our equivalence result allows us to dispense with this third Riccati equation entirely.

5.3 Equivalence with Minimum Entropy $\mathcal{H}_\infty$ Control

We begin with our main equivalence result.

Theorem 5.3.1 *With definitions and assumptions as in Section 5.2, for any (P,γ)-admissible closed-loop H, the entropy equals the auxiliary cost. That is,*

$$I(H;\gamma;\infty) = J(H;\gamma).$$

Hence, the Minimum Entropy $\mathcal{H}_\infty$ Control Problem (Problem 3.2.6) and the Combined $\mathcal{H}_\infty$/LQG Problem (Problem 5.2.2) are equivalent.

We defer the proof of Theorem 5.3.1 until after the next lemma, which gives an evaluation of the entropy of H in terms of its state-space matrices and the solution to an associated Riccati equation.

Lemma 5.3.2 *Let $H \in \mathcal{RH}_\infty$ with $\|H\|_\infty < \gamma$ and*

$$H = \left[\begin{array}{c|c} \tilde{A} & \tilde{B} \\ \hline \tilde{C} & 0 \end{array}\right].$$

Then its entropy is given by

$$I(H;\gamma;\infty) = \text{trace}[Q_s\tilde{C}^T\tilde{C}] \tag{5.5}$$

where $Q_s = Q_s^T \geq 0$ is the stabilizing solution to the algebraic Riccati equation

$$0 = \tilde{A}Q + Q\tilde{A}^T + \gamma^{-2}Q\tilde{C}^T\tilde{C}Q + \tilde{B}\tilde{B}^T. \tag{5.6}$$

Proof The assumption that $H \in \mathcal{RH}_\infty$, with $\|H\|_\infty < \gamma$, implies that

$$I - \gamma^{-2}H(j\omega)H^*(j\omega) > 0, \quad \forall\omega \in \mathbb{R}\cup\{\infty\}.$$

This guarantees the existence of a spectral factor N such that

$$I - \gamma^{-2}HH^* = NN^*, \quad \text{where } N^{\pm1} \in \mathcal{RH}_\infty.$$

It is easily verified that a state-space realization of N is (see e.g., [16, 60])

$$N = \left[\begin{array}{c|c} \tilde{A} & -\gamma^{-2}Q_s\tilde{C}^T \\ \hline \tilde{C} & I \end{array}\right],$$

where $Q_s = Q_s^T \geq 0$ is the stabilizing solution to the algebraic Riccati equation

$$0 = \tilde{A}Q + Q\tilde{A}^T + \gamma^{-2}Q\tilde{C}^T\tilde{C}Q + \tilde{B}\tilde{B}^T. \tag{5.7}$$

Using the fact that

$$\ln|\det(I - \gamma^{-2}H(j\omega)H^*(j\omega))| = \ln|\det(N(j\omega)N^*(j\omega))| = \ln|\det(N^*(j\omega)N(j\omega))|$$

we find, from the definition of entropy in Definition 2.2.2 and from Proposition 2.3.1(vii), that

$$\begin{aligned} I(H;\gamma;\infty) &= I(H^*;\gamma;\infty) \\ &= \lim_{s_0\to\infty}\left\{-\frac{\gamma^2}{2\pi}\int_{-\infty}^{\infty}\ln|\det(N^*(j\omega)N(j\omega))|\left[\frac{s_0^2}{s_0^2+\omega^2}\right]d\omega\right\}. \end{aligned} \tag{5.8}$$

We may immediately apply Lemma A.4.1 of Appendix A to get

$$I(H;\gamma;\infty) = \text{trace}[Q_s\tilde{C}^T\tilde{C}], \tag{5.9}$$

as claimed. □

Proof of Theorem 5.3.1 By definition (see (5.4) and (5.3)) the auxiliary cost is

$$J(H;\gamma) = \text{trace}[Q_s\tilde{C}^T\tilde{C}], \tag{5.10}$$

where $Q_s = Q_s^T \geq 0$ is the stabilizing solution to the algebraic Riccati equation

$$0 = \tilde{A}Q + Q\tilde{A}^T + \gamma^{-2}Q\tilde{C}^T\tilde{C}Q + \tilde{B}\tilde{B}^T. \tag{5.11}$$

A comparison with the evaluation of $I(H;\gamma;\infty)$ in Lemma 5.3.2 (compare (5.10) and (5.11) with (5.5) and (5.6)) establishes the result. □

Remark 5.3.3 In the definition of auxiliary cost given in [8], it is not mentioned that the *stabilizing* solution Q_s of the Riccati equation (5.3) should be used. However, it may be shown using the results of [60, 62] (taking care over the sign convention), that $Q \geq Q_s$, where Q is any solution to (5.3). Therefore, no other solution to (5.3) gives a smaller auxiliary cost than Q_s does, justifying our insistence on taking the stabilizing solution of (5.3) to evaluate $J(H;\gamma)$.

Section 3.7 states that recovery of the LQG problem associated with P is achieved by relaxing the $\mathcal{H}_\infty$-constraint completely (by allowing $\gamma \to \infty$ in the Minimum Entropy $\mathcal{H}_\infty$ Control Problem or equivalently in the Combined $\mathcal{H}_\infty$/LQG Problem). Then the entropy and the auxiliary cost both become the LQG cost: $I(H;\infty;\infty) = J(H;\infty) = C(H)$.

5.4 Solution and Equivalence in State-Space

In the previous section we proved the equivalence of the Minimum Entropy $\mathcal{H}_\infty$ Control Problem and the Combined $\mathcal{H}_\infty$/LQG Problem. In this section we expand on this by explicitly showing the equivalence of the state-space solutions of the two problems.

The equivalence result also allows us to use the solution to the Minimum Entropy $\mathcal{H}_\infty$ Control Problem to simplify the solution to the Combined $\mathcal{H}_\infty$/LQG Problem. We have, from Chapter 3, that the minimum entropy $\mathcal{H}_\infty$ controller and the minimum value of the entropy may be written in terms of the stabilizing solutions, denoted X_∞ and Y_∞, to two algebraic Riccati equations. The solution of the Combined $\mathcal{H}_\infty$/LQG Problem in [8] expresses the controller in terms of X_∞ and Y_∞ in the same way, but the minimum value of the auxiliary cost requires the solution, $\bar{Q}$, to a *third* algebraic Riccati equation coupled to the other two. Our equivalence result allows us to reduce the expression for the minimum value of the auxiliary cost to its equivalent, and simpler, expression as the minimum value of the entropy. This makes $\bar{Q}$ unnecessary. Before showing this, we can give a state-space realization of the controller.

Proposition 5.4.1 *The unique controller which solves the Minimum Entropy $\mathcal{H}_\infty$ Control Problem subject to the assumptions in Section 5.2 (equivalently, the controller which solves the Combined $\mathcal{H}_\infty$/LQG Problem as stated in Section 5.2) has a state-space realization*

$$K_{ME\infty} = \left[\begin{array}{c|c} A + Y_\infty(\gamma^{-2}C_1^TC_1 - C_2^TC_2) - B_2B_2^TX_\infty Z_\infty & Y_\infty C_2^T \\ \hline -B_2^TX_\infty Z_\infty & 0 \end{array}\right],$$

where $X_\infty = X_\infty^T \geq 0$, $Y_\infty = Y_\infty^T \geq 0$ are the stabilizing solutions to

$$X_\infty A + A^TX_\infty + C_1^TC_1 + X_\infty(\gamma^{-2}B_1B_1^T - B_2B_2^T)X_\infty = 0 \tag{5.12}$$

and

$$Y_\infty A^T + AY_\infty + B_1B_1^T + Y_\infty(\gamma^{-2}C_1^TC_1 - C_2^TC_2)Y_\infty = 0, \tag{5.13}$$

respectively, and where

$$Z_\infty := (I - \gamma^{-2} Y_\infty X_\infty)^{-1}.$$

In saying the stabilizing solutions, we mean the solutions X_∞ and Y_∞ such that

$$A + (\gamma^{-2} B_1 B_1^T - B_2 B_2^T) X_\infty \qquad \textit{is asymptotically stable,} \tag{5.14}$$

and

$$A + Y_\infty(\gamma^{-2} C_1^T C_1 - C_2^T C_2) \qquad \textit{is asymptotically stable.} \tag{5.15}$$

Proof The proposition follows in a straightforward way from the formulae given in Chapter 3 (Remark 3.5.6 in particular), on making use of the assumptions of Section 5.2. Alternatively, the method in [8] involves assuming the controller structure (with n states) and treating the problem as one of constrained optimization, yielding coupled equations which must be satisfied by the controller state-space matrices. Some straightforward manipulations lead to the above controller. □

The resulting closed-loop is given as usual by $H_{ME\infty} = \mathcal{F}(P, K_{ME\infty})$. State-space formulae for the minimum value of the entropy and the minimum value of the auxiliary cost can now be stated. These formulae are taken directly from Theorem 3.5.1 and [8], with certain simplifications implied by the assumptions of Section 5.2.

Proposition 5.4.2 *The minimum value of the entropy is given by*

$$I(H_{ME\infty}; \gamma; \infty) = \operatorname{trace}[X_\infty B_1 B_1^T + X_\infty Z_\infty Y_\infty X_\infty B_2 B_2^T]. \tag{5.16}$$

Proposition 5.4.3 ([8]) *The minimum value of the auxiliary cost is given by*

$$J(H_{ME\infty}; \gamma) = \operatorname{trace}[Y_\infty C_1^T C_1 + \bar{Q}_s (C_1^T C_1 + X_\infty Z_\infty B_2 B_2^T Z_\infty^T X_\infty)], \tag{5.17}$$

where $\bar{Q}_s = \bar{Q}_s^T \geq 0$ is the stabilizing solution to the algebraic Riccati equation

$$0 = \bar{A}\bar{Q} + \bar{Q}\bar{A}^T + \gamma^{-2}\bar{Q}\bar{R}\bar{Q} + Y_\infty C_2^T C_2 Y_\infty, \tag{5.18}$$

and

$$\begin{aligned} \bar{A} &:= A - B_2 B_2^T X_\infty Z_\infty + \gamma^{-2} Y_\infty C_1^T C_1, \\ \bar{R} &:= C_1^T C_1 + X_\infty Z_\infty B_2 B_2^T Z_\infty^T X_\infty. \end{aligned} \tag{5.19}$$

In saying the stabilizing solution, we mean the solution $\bar{Q}_s$ such that

$$\bar{A} + \gamma^{-2} \bar{Q}_s \bar{R}$$

is asymptotically stable.

Remark 5.4.4 By Theorem 5.3.1, $I(H_{ME\infty}; \gamma; \infty) = J(H_{ME\infty}; \gamma)$, which implies equality of (5.16) and (5.17). Therefore, we never need to solve equation (5.18) for $\bar{Q}_s$ to find $J(H_{ME\infty}; \gamma)$, as we can always use (5.16) instead, which only needs X_∞ and Y_∞. In this way, Theorem 5.3.1 and Proposition 5.4.2 allow us to *discard Proposition 5.4.3 as redundant.*

It is the purpose of the remainder of this section to clarify the nature of the redundancy of Proposition 5.4.3. We do this by firstly deriving some properties of the solutions to (5.18). We then use these properties to prove that (5.17) equals (5.16), and along the way see why it is the stabilizing solution to (5.18) which leads to the minimum value of the auxiliary cost.

In the interests of brevity and clarity, we make two mild simplifying assumptions as follows:

$$X_\infty > 0 \tag{5.20}$$

and

$$(\bar{A}^T, \gamma^{-1}\bar{R}^{1/2}) \quad \text{is controllable.} \tag{5.21}$$

Assumption (5.20) is also made in [8]; sufficient conditions for $X_\infty > 0$ may be found in [17]. Also, by [31], assumption (5.21) guarantees the existence of both a stabilizing solution $\bar{Q}_s$ and an antistabilizing solution $\bar{Q}_u$ to (5.18). Note that since the equivalence results of Section 5.3 and the final entropy formula (5.16) do *not* require either of these assumptions, it follows that these assumptions are not necessary—they merely streamline the presentation.

Lemma 5.4.5 *Assuming (5.20) and (5.21), the antistabilizing solution* $\bar{Q}_u = \bar{Q}_u^T > 0$ *to the algebraic Riccati equation (5.18) is*

$$\bar{Q}_u = \gamma^2 Z_\infty^{-1} X_\infty^{-1}.$$

Proof It is a simple matter to substitute $\bar{Q}_u = \gamma^2 Z_\infty^{-1} X_\infty^{-1}$ into equation (5.18) and use (5.12) and (5.13) to show that it is indeed a solution. That $\bar{Q}_u > 0$ follows from the assumption that $X_\infty > 0$ and the definition of Z_∞. It only remains to check that $\bar{Q}_u = \gamma^2 Z_\infty^{-1} X_\infty^{-1}$ is antistabilizing. In other words, we must show that

$$\bar{A}_u := \bar{A} + \gamma^{-2}\bar{Q}_u \bar{R}$$

is antistable. Substituting for $\bar{A}$, $\bar{R}$ and $\bar{Q}_u$,

$$\begin{aligned}
\bar{A}_u &= (A - B_2B_2^T X_\infty Z_\infty + \gamma^{-2} Y_\infty C_1^T C_1) + Z_\infty^{-1} X_\infty^{-1}(C_1^T C_1 + X_\infty Z_\infty B_2 B_2^T Z_\infty^T X_\infty) \\
&= A + X_\infty^{-1} C_1^T C_1 \\
&= X_\infty^{-1}(X_\infty A + C_1^T C_1) \\
&= -X_\infty^{-1}(A^T + X_\infty(\gamma^{-2}B_1B_1^T - B_2B_2^T))X_\infty,
\end{aligned} \tag{5.22}$$

where (5.12) has been used to obtain the final equality. Therefore $\bar{A}_u$ and $-(A + (\gamma^{-2}B_1B_1^T - B_2B_2^T)X_\infty)$ are similar matrices. But $A + (\gamma^{-2}B_1B_1^T - B_2B_2^T)X_\infty$ is asymptotically stable by (5.14). Hence $\bar{A}_u$ is antistable, as required. □

The solution $\gamma^2 Z_\infty^{-1} X_\infty^{-1}$ has also been obtained (independently) by Bernstein and Haddad. In order to deduce what we need to know about the stabilizing solution, based on our knowledge of this antistabilizing solution, we will need the following lemma. It is convenient to define the 'gap'

$$\Delta := \bar{Q}_u - \bar{Q}_s. \tag{5.23}$$

Lemma 5.4.6 *Assuming (5.20) and (5.21), we have that*

(i) Δ *is the stabilizing solution of the algebraic Riccati equation*

$$\bar{A}_u\Delta + \Delta\bar{A}_u^T - \gamma^{-2}\Delta\bar{R}\Delta = 0. \tag{5.24}$$

(ii) $\Delta > 0$.

Proof *Part (i)* That Δ satisfies (5.24) is easily verified using the definitions of $\bar{A}_u$, Δ and $\bar{R}$, and equation (5.18). To see that Δ is stabilizing, we note that

$$\begin{aligned}\bar{A}_u - \gamma^{-2}\Delta\bar{R} &= (\bar{A} + \gamma^{-2}\bar{Q}_u\bar{R}) - \gamma^{-2}(\bar{Q}_u - \bar{Q}_s)\bar{R} \\ &= \bar{A} + \gamma^{-2}\bar{Q}_s\bar{R},\end{aligned}$$

which is asymptotically stable by construction.

Part (ii) It is a consequence of assumption (5.21) that $(\bar{A}_u^T, \gamma^{-1}\bar{R}^{1/2})$ is controllable. This, together with the stability of $-\bar{A}_u^T$ and a well-known theorem on Lyapunov equations [26, Theorem 3.3], implies that

$$\Psi(-\bar{A}_u) + (-\bar{A}_u^T)\Psi + \gamma^{-2}\bar{R} = 0 \tag{5.25}$$

has a unique solution $\Psi > 0$. Comparing $\Psi^{-1}(5.25)\Psi^{-1}$ with (5.24), it follows that $\Psi^{-1} = \Delta$, so $\Delta > 0$ as claimed. □

Remark 5.4.7 If $\bar{Q}$ is any solution to (5.18), then

$$\bar{Q}_u \geq \bar{Q} \geq \bar{Q}_s \geq 0 \tag{5.26}$$

follows easily from [60, 62]. This ordering of solutions shows clearly that only the *stabilizing* solution, $\bar{Q}_s$, can lead to the *minimum* value of the auxiliary cost. A glance at (5.17), (5.19) and (5.23), together with (5.26) tells us that using the antistabilizing solution $\bar{Q}_u$, rather than the stabilizing solution $\bar{Q}_s$, leads to a cost which is exactly trace$[\Delta\bar{R}]$ bigger than the minimum.

Theorem 5.4.8 *The expression for the minimum value of the auxiliary cost in (5.17) equals the expression for the minimum value of the entropy in (5.16). That is,*

$$\begin{aligned}J(H_{ME\infty};\gamma) &= \text{trace}[Y_\infty C_1^T C_1 + \bar{Q}_s(C_1^T C_1 + X_\infty Z_\infty B_2 B_2^T Z_\infty^T X_\infty)] \\ &= \text{trace}[X_\infty B_1 B_1^T + X_\infty Z_\infty Y_\infty X_\infty B_2 B_2^T] = I(H_{ME\infty};\gamma;\infty),\end{aligned}$$

and hence we may dispense with Proposition 5.4.3 in favour of Proposition 5.4.2

Proof The conclusion of the theorem is just that of Theorem 5.3.1. What is of interest here is the direct proof of equality, by working only with the formulae and their associated Riccati equations. The LHS is just

$$\begin{aligned} LHS &= \text{trace}[Y_\infty C_1^T C_1 + \bar{Q}_s \bar{R}] \\ &= \text{trace}[Y_\infty C_1^T C_1 + \bar{Q}_u \bar{R}] - \text{trace}[\Delta \bar{R}] \\ &=: \quad (a) \quad - \quad (b). \end{aligned}$$

Using Lemma 5.4.5 to put $\bar{Q}_u = \gamma^2 Z_\infty^{-1} X_\infty^{-1}$, we get

$$\begin{aligned} (a) &= \text{trace}[Y_\infty C_1^T C_1 + \gamma^2 X_\infty^{-1} C_1^T C_1 - Y_\infty C_1^T C_1 + \gamma^2 B_2 B_2^T Z_\infty^T X_\infty] \\ &= \text{trace}[\gamma^2 X_\infty^{-1} C_1^T C_1 + \gamma^2 B_2 B_2^T Z_\infty^T X_\infty]. \end{aligned}$$

From Lemma 5.4.6,

$$\begin{aligned} (b) &= \gamma^2 \text{ trace}[\bar{A}_u + \Delta \bar{A}_u^T \Delta^{-1}] \\ &= 2\gamma^2 \text{ trace}[\bar{A}_u] \\ &= 2\gamma^2 \text{ trace}[A + X_\infty^{-1} C_1^T C_1] \\ &= \gamma^2 \text{ trace}[X_\infty^{-1}(A^T X_\infty + X_\infty A + C_1^T C_1) + X_\infty^{-1} C_1^T C_1] \\ &= -\text{ trace}[B_1 B_1^T X_\infty - \gamma^2 B_2 B_2^T X_\infty - \gamma^2 X_\infty^{-1} C_1^T C_1], \end{aligned}$$

where we have used (5.22) to obtain the third equality, and X_∞^{-1}(5.12) to obtain the final equality. Hence,

$$\begin{aligned} LHS &= (a) - (b) \\ &= \text{trace}[\gamma^2 X_\infty^{-1} C_1^T C_1 + \gamma^2 B_2 B_2^T Z_\infty^T X_\infty] \\ &\qquad\qquad + \text{trace}[B_1 B_1^T X_\infty - \gamma^2 B_2 B_2^T X_\infty - \gamma^2 X_\infty^{-1} C_1^T C_1] \\ &= \text{trace}[X_\infty B_1 B_1^T + \gamma^2 (Z_\infty^T - I) X_\infty B_2 B_2^T] \\ &= RHS. \end{aligned}$$

□

Chapter 6

Relations to Risk-Sensitive LQG Control

6.1 Introduction

In previous chapters we have seen the importance of the entropy at infinity and have identified it as an auxiliary LQG cost. The aim of this chapter is to further our understanding of the entropy. We shall do this by relating the Minimum Entropy $\mathcal{H}_\infty$ Control Problem to an apparently unrelated problem, the Risk-Sensitive LQG Control Problem. Risk-sensitive LQG control is a generalization of usual LQG control to include an *exponential-of-quadratic cost* (the LEQG cost). The equivalence of the minimum entropy $\mathcal{H}_\infty$ controller and the steady-state version of the risk-sensitive optimal controller was first established in [28], for discrete-time systems. It is interesting that two conceptually quite different control methodologies give the same results and this can be considered as a bonus to both approaches; each performing well with respect to a criterion that had not been explicitly considered. We hence have a stochastic interpretation of the minimum entropy $\mathcal{H}_\infty$ controller and an interpretation of the LEQG controller with respect to robust stability in the presence of plant uncertainty and signal gain to worst case disturbances.

The purpose of the present chapter is to give the continuous-time version of the equivalence. Our analysis follows [27]. Note that, in keeping with the remainder of this monograph, the finite-dimensional case is considered. In [27], the result is extended to cover a class of infinite-dimensional systems, too.

6.2 The Risk-Sensitive LQG Control Problem

The study of the LEQG problem was begun in [34] where the solution for perfect state observation was derived. Then in [57], the imperfect output observation case was solved in discrete-time, by developing a risk-sensitive certainty equivalence principle. The demonstration in [28] of the equivalence between the LEQG and entropy criteria prompted [58], where it is shown that the Hamiltonian methods developed in [59] can be applied to give a direct derivation of the entropy minimizing property of the LEQG controller. Although this latter paper only considers the discrete-time problem with perfect state observations, the techniques could be adapted to other situations. The continuous-time case with imperfect output observation, finite time horizon and state-space, and possibly time-varying system model, was solved in [7].

The set-up will be our standard one, as shown in Figure 3.1; $H = \mathcal{F}(P, K)$ is taken to be a stabilized closed-loop. The input $w(t)$ is taken to be zero mean Gaussian white noise with spectrum equal to the identity. Then the standard finite-time LQG control problem is to minimize the finite-time LQG cost (see Remark 2.4.2)

$$C_T(H) = \mathbf{E}\{W_T\},$$

over all stabilizing controllers, where

$$W_T = \frac{1}{2T}\int_{-T}^{T} z^T(t)z(t)dt. \tag{6.1}$$

The *Risk-Sensitive LQG Control Problem* is to minimize the modulus of the *exponential-of-quadratic cost*

$$\Omega_T(\theta) := -(\theta T)^{-1} \ln \mathbf{E}\{\exp(-\theta T W_T)\} \tag{6.2}$$

over all stabilizing controllers.

The scalar θ is the *risk-sensitivity parameter*. The standard LQG control problem is recovered when $\theta \to 0$ (risk-neutral); $\theta > 0$ is risk-seeking and $\theta < 0$ is risk-averse. To help interpret this, following [57, p765–6], if $\theta T \mathrm{Var}\{W_T\}$ is small then a power series expansion gives

$$\Omega_T(\theta) \approx \mathbf{E}\{W_T\} - (\theta T/2)\mathrm{Var}\{W_T\}.$$

If $\theta > 0$ then, for a given value of $\mathbf{E}\{W_T\}$, any statistical variation in W_T improves the value of the exponential-of-quadratic cost. This is an optimistic situation where incidence of favourable values of W_T (i.e., less than $\mathbf{E}\{W_T\}$) is taken to be more significant than the incidence of unfavourable values of W_T (i.e., greater than $\mathbf{E}\{W_T\}$). If $\theta < 0$ then the situation is reversed and the controller becomes pessimistic, in that incidence of unfavourable values of W_T is taken to be more significant than incidence of favourable values of W_T. The risk-neutral case ($\theta \to 0$) lies at the border of optimism and pessimism: occurrences of favourable or unfavourable excursions of W_T are assumed to be equally significant.

6.3 Equivalence with Minimum Entropy $\mathcal{H}_\infty$ Control

We shall be concerned with the risk-averse case so we set $\theta = -\gamma^{-2} < 0$. The LEQG cost becomes

$$J_T(\gamma) := \Omega_T(\theta) = (\gamma^2/T) \ln \mathbf{E}\{\exp(\gamma^{-2} T W_T)\}, \tag{6.3}$$

where W_T is as in (6.1) The equivalence between the Risk-Sensitive LQG Control Problem and the Minimum Entropy $\mathcal{H}_\infty$ Control Problem is then a consequence of the following result.

Proposition 6.3.1 ([27]) *Let $H \in \mathcal{RH}_\infty$, and let z be the output of this system when driven by a Gaussian white noise input w with spectrum equal to the identity. Then the LEQG cost in the risk-averse case ($\theta = -\gamma^{-2} < 0$), and the entropy, satisfy*

$$J_T(\gamma) = \begin{cases} I(H;\gamma;\infty) + O(1/T) & \text{if } \|H\|_\infty < \gamma, \\ \infty & \text{if } \|H\|_\infty > \gamma \text{ and } T \text{ is sufficiently large.} \end{cases}$$

Thus in the steady-state case ($T \to \infty$),

$$J_\infty(\gamma) = I(H;\gamma;\infty) \quad \text{if } \|H\|_\infty < \gamma.$$

That is, for any (P,γ)-admissible closed-loop H, the entropy $I(H;\gamma;\infty)$ equals the steady-state LEQG cost $J_\infty(\gamma)$.

Proof This proof follows [27] and is included for completeness. Assume that the closed-loop system H satisfies the following stochastic differential equation,

$$\begin{aligned} dx_t &= Ax_t + Bdv_t \\ z_t &= Cx_t \end{aligned}$$

where v_t is a normalized Brownian motion. To evaluate $J_T(\gamma)$ consider the value function conditioned on the state at time t,

$$f(t,x) = \mathbf{E}|_{x_t=x}\left\{\exp\left(\frac{\gamma^{-2}}{2}\int_t^T z_s^T z_s\, ds\right)\right\}.$$

A standard calculation (see for example [13, page 195]) now gives

$$\begin{aligned} f(t,x) &= \exp\left\{(\gamma^{-2}/2)x^T C^T C x dt + o(dt)\right\} \\ &\times\left\{f(t,x) + \frac{\partial f}{\partial x}Ax\, dt + (1/2)\text{trace}\left[\frac{\partial^2 f}{\partial x^2}BB^T dt\right] + \frac{\partial f}{\partial t}dt + o(dt)\right\} \end{aligned}$$

which implies that

$$\frac{\partial f}{\partial t} + \frac{\partial f}{\partial x}Ax + (\gamma^{-2}/2)x^T C^T C x f(t,x) + (1/2)\text{trace}\left[\frac{\partial^2 f}{\partial x^2}BB^T\right] = 0. \tag{6.4}$$

Following [34], this is satisfied by

$$f(t,x) = g(t)\exp\left\{(\gamma^{-2}/2)x^T S(t)x\right\}$$

where S and g satisfy

$$-\dot{S}(t) = C^T C + S(t)A + A^T S(t) + \gamma^{-2}S(t)BB^T S(t), \quad S(T) = 0 \tag{6.5}$$

$$-\dot{g}(t) = (\gamma^{-2}/2)g(t)\ \text{trace}[S(t)BB^T], \quad g(T) = 1. \tag{6.6}$$

This is verified to be a solution by noting that,

$$\begin{aligned} \frac{\partial f}{\partial t} &= \left[\dot{g} + (\gamma^{-2}/2)gx^T\dot{S}x\right]\exp\{(\gamma^{-2}/2)x^T Sx\} \\ \frac{\partial f}{\partial x} &= \left[\gamma^{-2}gx^T S\right]\exp\{(\gamma^{-2}/2)x^T Sx\} \\ \frac{\partial^2 f}{\partial x^2} &= g\gamma^{-2}\left[S + \gamma^{-2}Sxx^T S\right]\exp\{(\gamma^{-2}/2)x^T Sx\}. \end{aligned}$$

Substituting into (6.4) gives,

$$\begin{aligned} &\exp\{-(\gamma^{-2}/2)x^T Sx\} \times \text{left-hand side of (6.4)} \\ &= \dot{g} + (\gamma^{-2}/2)gx^T\dot{S}x + \gamma^{-2}gx^T SAx + (\gamma^{-2}/2)gx^T C^T Cx \\ &\quad + (\gamma^{-2}/2)g\ \text{trace}\left[(S + \gamma^{-2}Sxx^T S)BB^T\right] \\ &= \dot{g} + (\gamma^{-2}/2)g\ \text{trace}\left[SBB^T\right] \\ &\quad + (\gamma^{-2}/2)gx^T(\dot{S} + C^T C + SA + A^T S + \gamma^{-2}SBB^T S)x \\ &= 0 \text{ by (6.5) and (6.6).} \end{aligned}$$

In order to evaluate $J_T(\gamma)$ in (6.3) under the assumption that the state $x_{-T} \sim N(0, P)$, consider

$$\begin{aligned}
&\mathbf{E}\left\{\exp\{\gamma^{-2}TW_T\}\right\} \\
&= \int\cdots\int \mathbf{E}|_{x_{-T}=x}\left\{\exp\{\gamma^{-2}TW_T\}\right\}\left\{\frac{\exp\{-(1/2)x^TP^{-1}x\}}{\sqrt{(2\pi)^n\det(P)}}\right\}dx \\
&= \int\cdots\int g(-T)\exp\{(\gamma^{-2}/2)x^TS(-T)x\}\left\{\frac{\exp\{-(1/2)x^TP^{-1}x\}}{\sqrt{(2\pi)^n\det(P)}}\right\}dx \\
&= \frac{g(-T)}{\sqrt{\det(I-\gamma^{-2}PS(-T))}}
\end{aligned}$$

and hence

$$J_T(\gamma) = (\gamma^2/T)\ln(g(-T)) - (\gamma^2/2T)\ln\det(I - \gamma^{-2}PS(-T)). \tag{6.7}$$

The steady state case is obtained by taking the limit as $T \to \infty$. Firstly observe that $S(t) \to S_\infty$ as $T - t \to \infty$ where S_∞ is the stabilizing solution to the algebraic Riccati equation corresponding to (6.5), i.e.,

$$C^TC + S_\infty A + A^TS_\infty + \gamma^{-2}S_\infty BB^TS_\infty = 0, \quad \mathrm{Re}\{\lambda_i\{A + \gamma^{-2}BB^TS_\infty\}\} < 0$$

and S_∞ exists if and only if $\|H\|_\infty < \gamma$. This is a standard result on the Riccati differential equation and can be found in Remark 21 in [60]. Now let $h(t) := \ln(g(t))$ then

$$\dot{h} = -(\gamma^{-2}/2)\mathrm{trace}[S(t)BB^T]$$

and

$$\lim_{T\to\infty}\{h(-T)/T\} = \gamma^{-2}\mathrm{trace}[S_\infty BB^T]$$

and hence on substituting into (6.7), the LEQG cost is given by

$$J_T(\gamma) = \mathrm{trace}[S_\infty BB^T] + O(1/T).$$

Finally this expression can be related to the entropy integral by using Lemma 5.3.2 on H^T. We get, after noting Proposition 2.3.1(vii), that

$$I(H;\gamma;\infty) = I(H^T;\gamma;\infty) = \mathrm{trace}[S_\infty BB^T],$$

and Proposition 6.3.1 follows. □

As a final remark, note that as $\theta = -\gamma^{-2}$ is made more negative there comes a point where all controllers give infinite exponential-of-quadratic cost. This corresponds to maximal risk-aversion or equivalently, to $\mathcal{H}_\infty$-optimality (since $\|H\|_\infty$ is then minimized over all stabilizing controllers).

Chapter 7

The Normalized $\mathcal{H}_\infty$ Control Problem

7.1 Introduction

In Section 3.7 we found that relaxing the $\mathcal{H}_\infty$-norm bound in any Minimum Entropy $\mathcal{H}_\infty$ Control Problem recovers an LQG control problem. Here we shall reverse this argument. We begin with the Normalized LQG Control Problem of [35] and obtain its associated Minimum Entropy $\mathcal{H}_\infty$ Control Problem, which we call the Normalized $\mathcal{H}_\infty$ Control Problem. Proposition 3.6.1 implies that in imposing the minimum entropy/$\mathcal{H}_\infty$ criterion to obtain the Normalized $\mathcal{H}_\infty$ Problem, we obtain robustness at the expense of the (upper bound on the) LQG cost. We give a simple numerical example to make the ideas concrete and to illustrate that the improvements in the robustness guarantees may be large relative to the degradation in LQG cost.

7.2 The Normalized LQG Problem

The Normalized LQG Problem of [35] is based on a system

$$\begin{aligned} \dot{x} &= Ax + Bw_1 + Bu \\ z_1 &= Cx, \qquad z_2 = u \\ y &= Cx + w_2, \end{aligned}$$

where w_1 and w_2 are zero mean Gaussian white noise signals, each with a spectrum equal to the identity, and the given plant $G = (A, B, C)$ is stabilizable and detectable. Put $z := [z_1^T\ z_2^T]^T$ and $w := [w_1^T\ w_2^T]^T$. The LQG cost (see Definition 2.4.1) is just

$$C(H) = \lim_{T\to\infty} \mathbf{E}\left\{\frac{1}{2T}\int_{-T}^{T} x^T(t)C^TCx(t) + u^T(t)u(t)\,dt\right\}.$$

See Figure 7.1 for a block diagram of the closed-loop system.

It is easily seen that the Normalized LQG Problem has a standard plant (in the sense of Section 3.2 and Figure 3.1) given by

$$P = \left[\begin{array}{c|cc} A & \begin{bmatrix} B & 0 \end{bmatrix} & \begin{bmatrix} B \end{bmatrix} \\ \hline \begin{bmatrix} C \\ 0 \end{bmatrix} & \begin{bmatrix} 0 & 0 \\ 0 & 0 \end{bmatrix} & \begin{bmatrix} 0 \\ I \end{bmatrix} \\ \begin{bmatrix} C \end{bmatrix} & \begin{bmatrix} 0 & I \end{bmatrix} & \begin{bmatrix} 0 \end{bmatrix} \end{array}\right] \tag{7.1}$$

Some simple manipulations show that the closed-loop transfer function from w to z is

$$H = \mathcal{F}(P, K) = \begin{bmatrix} SG & SGK \\ KSG & KS \end{bmatrix} \tag{7.2}$$

where $S := (I - GK)^{-1}$. We recognise this system to be the one used to analyse internal stability [56, p101]. Note that because of the form of P and Remark 3.2.2, a controller K stabilizes P if and only if it stabilizes $P_{22} = G$.

The solution to the Normalized LQG Problem follows immediately from the state-space formulae of Section 3.7, or from standard references [2, 38, 35].

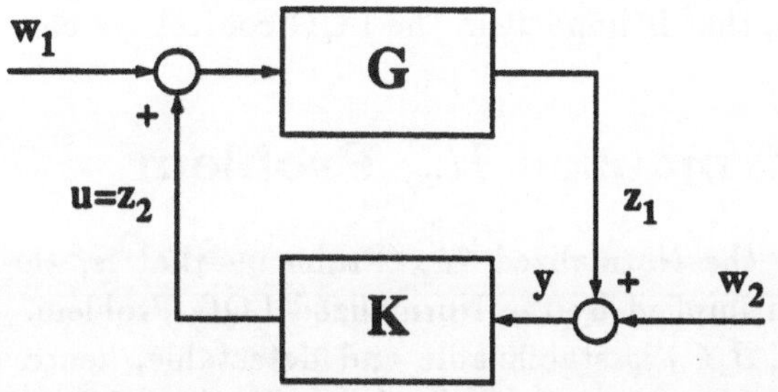

Figure 7.1: Block diagram for the Normalized Problems

Proposition 7.2.1 (Solution to the Normalized LQG Problem)
Let $G = (A, B, C)$ be stabilizable and detectable. Then there exists a unique positive semidefinite stabilizing solution $X_2 = X_2^T$ of the control algebraic Riccati equation (CARE):

$$A^T X_2 + X_2 A - X_2 B B^T X_2 + C^T C = 0 \qquad CARE \tag{7.3}$$

and there also exists a unique positive semidefinite stabilizing solution $Y_2 = Y_2^T$ of the filter algebraic Riccati equation (FARE):

$$A Y_2 + Y_2 A^T - Y_2 C^T C Y_2 + B B^T = 0. \qquad FARE \tag{7.4}$$

The Normalized LQG Controller $K_{LQG} = (\hat{A}, \hat{B}, \hat{C})$ takes the form of an optimal observer

$$\dot{\hat{x}} = \underbrace{(A - Y_2 C^T C - B B^T X_2)}_{\hat{A}} \hat{x} + \underbrace{Y_2 C^T}_{\hat{B}} y \tag{7.5}$$

together with the optimal state-feedback

$$u = \underbrace{-B^T X_2}_{\hat{C}} \hat{x}. \tag{7.6}$$

The minimum value of the LQG cost is

$$C(H_{LQG}) = \text{trace}[B^T X_2 B + B^T X_2 Y_2 X_2 B], \tag{7.7}$$

where H_{LQG} is the LQG-optimal closed-loop (i.e., H in equation (7.2) with K_{LQG}).

Remark 7.2.2 If $G = (A, B, C)$ is actually minimal, rather than just stabilizable and detectable, the above proposition is unchanged except both X_2 and Y_2 are then positive definite rather than positive semidefinite.

Note that the first and second terms in (7.7) are associated with the control and filtering, respectively, that is implicit in the LQG control system.

7.3 The Normalized $\mathcal{H}_\infty$ Problem

Now let us examine the Normalized $\mathcal{H}_\infty$ Problem—that is, the Minimum Entropy $\mathcal{H}_\infty$ Control Problem implied by the Normalized LQG Problem. By assumption, the given plant $G = (A, B, C)$ is stabilizable and detectable, hence Assumption 3.2.1 is fulfilled. Also, it is easy to see that all of Assumptions 3.2.7 are fulfilled, so P as in (7.1) is indeed a legitimate standard plant for the Minimum Entropy $\mathcal{H}_\infty$ Control Problem. The block diagram is identical to that for the Normalized LQG Problem shown in Figure 7.1; the closed-loop transfer function H is given in (7.2). We noted in Section 7.2 that because of the form of P and Remark 3.2.2, a controller K stabilizes P if and only if it stabilizes G. Hence a controller K is (P,γ)-admissible if and only if K stabilizes G and $\|H\|_\infty < \gamma$. This allows us to use the results of Section 3.5 to state the solution to the Normalized $\mathcal{H}_\infty$ Problem. But firstly some remarks on robustness.

Remark 7.3.1 Each element of H has a robustness interpretation: SG corresponds to 'additive' uncertainty Δ_{11} on the controller K; SGK corresponds to 'output multiplicative' uncertainty Δ_{12} on the plant G; KSG corresponds to 'input multiplicative' uncertainty Δ_{21} on G; and KS corresponds to additive uncertainty Δ_{22} on G. Block diagrams illustrating these four uncertainty types Δ_{ij}, $i,j = 1,2$, are provided in Figure 7.2. (See [19] for a discussion of various uncertainty types). The $\mathcal{H}_\infty$-norm bound on H implies robust stability guarantees for each of the four uncertainty types, via the Small Gain Theorem. This follows from the simple observation that

$$\left\| \begin{bmatrix} SG & SGK \\ KSG & KS \end{bmatrix} \right\|_\infty < \gamma \Longrightarrow \begin{cases} \|SG\|_\infty < \gamma \\ \|SGK\|_\infty < \gamma \\ \|KSG\|_\infty < \gamma \\ \|KS\|_\infty < \gamma \end{cases} \tag{7.8}$$

Using the Small Gain Theorem as in Remark 2.5.2 implies that closed-loop stability is maintained for any one $\Delta_{ij} \in \mathcal{RH}_\infty$ satisfying

$$\|\Delta_{ij}\|_\infty \leq \gamma^{-1}, \qquad i,j = 1,2.$$

(Note that because K stabilizes G, we have $SG \in \mathcal{RH}_\infty$, $SGK \in \mathcal{RH}_\infty$, $KSG \in \mathcal{RH}_\infty$ and $KS \in \mathcal{RH}_\infty$.) If we use the Small Gain Theorem on a frequency-by-frequency

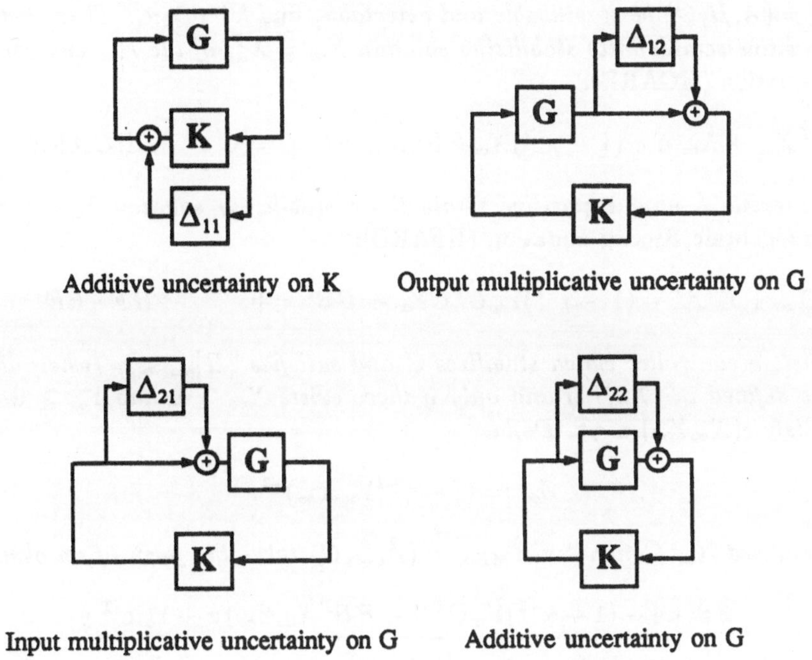

Figure 7.2: Uncertainty types Δ_{ij} covered by the Normalized $\mathcal{H}_\infty$ Problem

basis (as in Theorem 2.5.1) then a frequency-wise description of the tolerable Δ_{ij} is obtained: closed-loop stability is maintained for any one $\Delta_{ij} \in \mathcal{RH}_\infty$ satisfying

$$\begin{aligned}
\sigma_1\{\Delta_{11}(j\omega)\}\,\sigma_1\{S(j\omega)G(j\omega)\} &< 1 && \forall\omega \in \mathbb{R}\cup\{\infty\};\\
\sigma_1\{\Delta_{12}(j\omega)\}\,\sigma_1\{S(j\omega)G(j\omega)K(j\omega)\} &< 1 && \forall\omega \in \mathbb{R}\cup\{\infty\};\\
\sigma_1\{\Delta_{21}(j\omega)\}\,\sigma_1\{K(j\omega)S(j\omega)G(j\omega)\} &< 1 && \forall\omega \in \mathbb{R}\cup\{\infty\};\\
\sigma_1\{\Delta_{22}(j\omega)\}\,\sigma_1\{K(j\omega)S(j\omega)\} &< 1 && \forall\omega \in \mathbb{R}\cup\{\infty\}.
\end{aligned}$$

Remark 7.3.2 A bound on the sensitivity is obtained from the identity

$$S = I + SGK,$$

by taking the $\mathcal{H}_\infty$-norm. Use the triangle inequality together with the appropriate bound from (7.8) to give

$$\|S\|_\infty < 1 + \gamma.$$

Proposition 7.3.3 (Solution to the Normalized $\mathcal{H}_\infty$ Problem)
Let $G = (A, B, C)$ be stabilizable and detectable, and let $\gamma > \gamma_o$. Then there exists a unique positive semidefinite stabilizing solution $X_\infty = X_\infty^T$ of the $\mathcal{H}_\infty$ control algebraic Riccati equation (HCARE):

$$A^T X_\infty + X_\infty A - (1-\gamma^{-2}) X_\infty B B^T X_\infty + C^T C = 0 \qquad HCARE \qquad (7.9)$$

and there exists a unique positive semidefinite stabilizing solution $Y_\infty = Y_\infty^T$ of the $\mathcal{H}_\infty$ filter algebraic Riccati equation (HFARE):

$$A Y_\infty + Y_\infty A^T - (1-\gamma^{-2}) Y_\infty C^T C Y_\infty + B B^T = 0. \qquad HFARE \qquad (7.10)$$

There exists a controller which stabilizes G and satisfies $\|H\|_\infty < \gamma$ (where the closed-loop H is defined in (7.2)) if and only if there exists $X_\infty \geq 0$ and $Y_\infty \geq 0$ as above, which satisfy $\rho(X_\infty Y_\infty) < \gamma^2$. Define

$$Z_\infty := (I - \gamma^{-2} Y_\infty X_\infty)^{-1}. \qquad (7.11)$$

The Normalized $\mathcal{H}_\infty$ Controller $K_{ME\infty} = (\hat{A}, \hat{B}, \hat{C})$ takes the form of an observer

$$\dot{\hat{x}} = \underbrace{(A - (1-\gamma^{-2}) Y_\infty C^T C - B B^T X_\infty Z_\infty)}_{\hat{A}} \hat{x} + \underbrace{Y_\infty C^T}_{\hat{B}} y \qquad (7.12)$$

together with the state-feedback

$$u = \underbrace{-B^T X_\infty Z_\infty}_{\hat{C}} \hat{x}. \qquad (7.13)$$

The minimum value of the entropy is

$$I(H_{ME\infty}; \gamma; \infty) = \text{trace}[B^T X_\infty B + B^T X_\infty Z_\infty Y_\infty X_\infty B], \qquad (7.14)$$

where $H_{ME\infty}$ is the minimum entropy closed-loop (i.e., H in equation (7.2) with $K_{ME\infty}$).

Remark 7.3.4 If $G = (A, B, C)$ is actually minimal, rather than just stabilizable and detectable, the above proposition is unchanged except both X_∞ and Y_∞ are then positive definite rather than positive semidefinite.

To solve the $\mathcal{H}_\infty$-*optimal* problem for this plant, we apply the above proposition: there exists a stabilizing controller such that $\|H\|_\infty < \gamma$, if and only if there exists $X_\infty \geq 0$ and $Y_\infty \geq 0$ as above, which satisfy $\rho(X_\infty Y_\infty) < \gamma^2$. The optimal $\mathcal{H}_\infty$-norm γ_o is the infimal value of γ. In general a closed-form solution for γ_o is not available, but γ-iteration can be used to isolate γ_o to an arbitrary accuracy. The following remark and lemma are interesting here: Corollary 8.4.6 should also be consulted.

Remark 7.3.5 By definition

$$\gamma_o = \inf_K\{\|H\|_\infty \ : \ K \text{ stabilizes } G\},$$

where $H = \begin{bmatrix} SG & SGK \\ KSG & KS \end{bmatrix}$ and $S = (I-GK)^{-1}$. Consider the closely related problem

$$\hat{\gamma}_o := \inf_K\{\|\hat{H}\|_\infty \ : \ K \text{ stabilizes } G\},$$

where

$$\hat{H} := H + \begin{bmatrix} 0 & I \\ 0 & 0 \end{bmatrix} = \begin{bmatrix} SG & S \\ KSG & KS \end{bmatrix}. \tag{7.15}$$

This latter problem has an exact (non-iterative) solution [40, Remark 4.21] in terms of X_2 and Y_2, the stabilizing solutions to the CARE and FARE, respectively:

$$\hat{\gamma}_o = \sqrt{1+\lambda_1\{X_2Y_2\}} > 1.$$

But the triangle inequality applied to the $\mathcal{H}_\infty$-norm of (7.15) gives

$$\|\hat{H}\|_\infty - 1 \leq \|H\|_\infty \leq \|\hat{H}\|_\infty + 1.$$

Taking the infimum over all K which stabilize G gives

$$0 < \hat{\gamma}_o - 1 \leq \gamma_o \leq \hat{\gamma}_o + 1,$$

which provides (non-iterative) upper and lower bounds on γ_o.

Lemma 7.3.6 *Consider the Normalized $\mathcal{H}_\infty$ Problem for a given stabilizable and detectable system $G = (A, B, C)$—that is, consider the Minimum Entropy $\mathcal{H}_\infty$ Control Problem (Problem 3.2.6) for a standard plant P as given in (7.1). Then $\gamma_o < 1$ only if the given plant $G = (A, B, C)$ is asymptotically stable. Conversely, if the given plant $G = (A, B, C)$ is not asymptotically stable then $\gamma_o \geq 1$.*

Proof Suppose $\gamma_o < \gamma \leq 1$. Then (7.10) can be written as a Lyapunov equation:

$$0 = Y_\infty A^T + AY_\infty + [\alpha Y_\infty C^T \ \ B][\alpha Y_\infty C^T \ \ B]^T,$$

where $\alpha^2 := \gamma^{-2} - 1 \geq 0$. The pair $(A, [\alpha Y_\infty C^T \ \ B])$ is stabilizable because, by assumption, (A, B) is. This together with $Y_\infty \geq 0$ implies that A is asymptotically stable, by a standard result on Lyapunov equations [63, Lemma 12.2]. □

For convenience, we state the upper bounds on $\mathcal{H}_\infty$-norm and LQG cost provided by Proposition 3.6.1.

Proposition 7.3.7 *The $\mathcal{H}_\infty$-norm and the LQG cost of the minimum entropy closed loop $H_{ME\infty} = \mathcal{F}(P, K_{ME\infty})$, satisfy*

$$\|H_{ME\infty}\|_\infty < \gamma \tag{7.16}$$

$$C(H_{ME\infty}) \leq \ \operatorname{trace}[B^TX_\infty B + B^TX_\infty Z_\infty Y_\infty X_\infty B] \ . \tag{7.17}$$

The tradeoff of Theorem 3.6.2 between the $\mathcal{H}_\infty$ bound and LQG cost bound is then apparent from the discussion in Section 3.7: as the RHS of (7.16) increases (*resp.* decreases) so the RHS of (7.17) decreases (*resp.* increases) monotonically.

7.4 A Numerical Example

Let us demonstrate the results of the previous section on a simple numerical example, chosen for its ability to illustrate the tradeoffs rather than for physical interpretation. Take

$$G = \left[\begin{array}{c|c} A & B \\ \hline C & 0 \end{array}\right] = \left[\begin{array}{cc|c} 20 & -100 & 1 \\ 1 & 0 & 0 \\ \hline 1 & -0.1 & 0 \end{array}\right]$$

This has a right-half plane zero at 0.1 and two right-half plane poles at 10. Using γ-iteration gives $\gamma_0 = 40.87$ (the minimum value of $\|H\|_\infty$ i.e., this corresponds to the $\mathcal{H}_\infty$-optimal solution). Then, solving the X_2 and Y_2 equations, we find from (7.7) that $C(H_{LQG}) = 64 \times 10^3$ (the minimum value of the LQG cost i.e., this corresponds to the LQG solution). For any $\gamma > \gamma_o$, X_∞ and Y_∞ are found from (7.9) and (7.10), and the minimum entropy controller, $K_{ME\infty}$, is found using (7.12) and (7.13). Then the minimum entropy closed-loop is $H_{ME\infty} = \mathcal{F}(P, K_{ME\infty}) := (\tilde{A}, \tilde{B}, \tilde{C})$, say. The minimum value of the entropy is evaluated using (7.14). To calculate $C(H_{ME\infty})$, solve the Lyapunov equation for the controllability Gramian $\tilde{Q}$ of the minimum entropy closed-loop; then $C(H_{ME\infty}) = \mathrm{trace}[\tilde{Q}\tilde{C}^T\tilde{C}]$ as in (5.1). Calculation of $\|H_{ME\infty}\|_\infty$ can be performed to prespecified accuracy using the algorithm in [10]. These calculations were done for a number of values of γ, ranging over several orders of magnitude from close to γ_o. The results are illustrated in Figures 7.3-7.7.

Figure 7.3 is a plot of (7.16) as γ varies i.e., of the upper bound γ on $\|H_{ME\infty}\|_\infty$ and the actual value of $\|H_{ME\infty}\|_\infty$, against γ. Figure 7.4 is a plot of (7.17) as γ varies i.e., of the upper bound $I(H_{ME\infty};\gamma;\infty)$ on $C(H_{ME\infty})$ and the actual value of $C(H_{ME\infty})$, against γ. The graphs illustrate clearly the $\mathcal{H}_\infty$/LQG tradeoff of Theorem 3.6.2: the upper curve in Figure 7.3 *increases* with increasing γ, whilst the upper curve in Figure 7.4 *decreases* with increasing γ. In fact, we notice that the tradeoff is even stronger than this—the *achieved* values (i.e., the lower curves in Figures 7.3 and 7.4) exhibit the same behaviour as their upper bounds. See Chapter 9 for more on such monotonicity.

Notice that in Figure 7.3, as γ becomes large, $\|H_{ME\infty}\|_\infty$ tends fairly slowly towards $\|H_{LQG}\|_\infty \approx 80$, a number slightly less than twice γ_o. The variation with γ is more rapid in Figure 7.4. Both $C(H_{ME\infty})$ and its upper bound $I(H_{ME\infty};\gamma;\infty)$ decrease quickly with γ when γ is close to γ_o. So large improvements in the LQG properties can be obtained with only modest degradation of $\mathcal{H}_\infty$ properties. Although theory predicts that γ must be arbitrarily large before the Minimum Entropy $\mathcal{H}_\infty$ Control Problem problem reduces to the LQG problem, we see in Figure 7.4 that for γ as small as twice γ_o, the LQG cost and its bound are very close to their minimum (LQG-optimal) values. Furthermore, the upper bound provided by the entropy is quite tight, as we would expect from Theorem 2.4.4(ii).

Figure 7.5 shows the maximum singular value of the maximum entropy closed-loop as a function of frequency. Five curves are plotted, corresponding to five representative values of γ: $\gamma = 40.87$ ($\mathcal{H}_\infty$-optimal), 43, 45, 60 and ∞ (LQG-optimal). The rapid convergence (as γ increases) towards the LQG curve is clear, as is the fairly slow

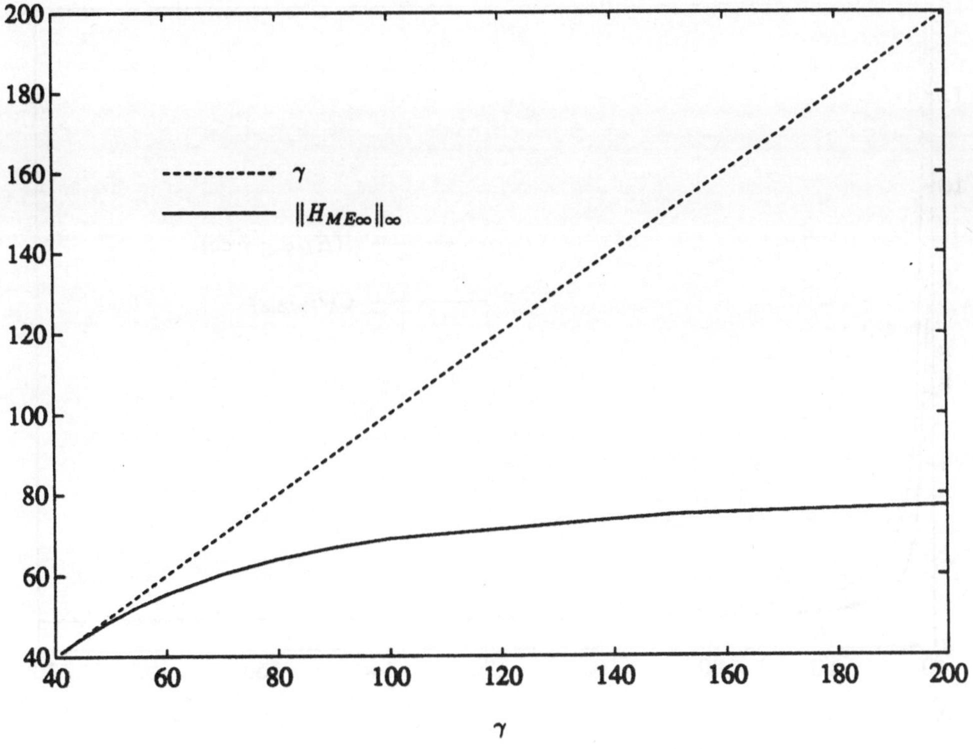

Figure 7.3: $\|H_{ME\infty}\|_\infty$ and its upper bound γ, against γ

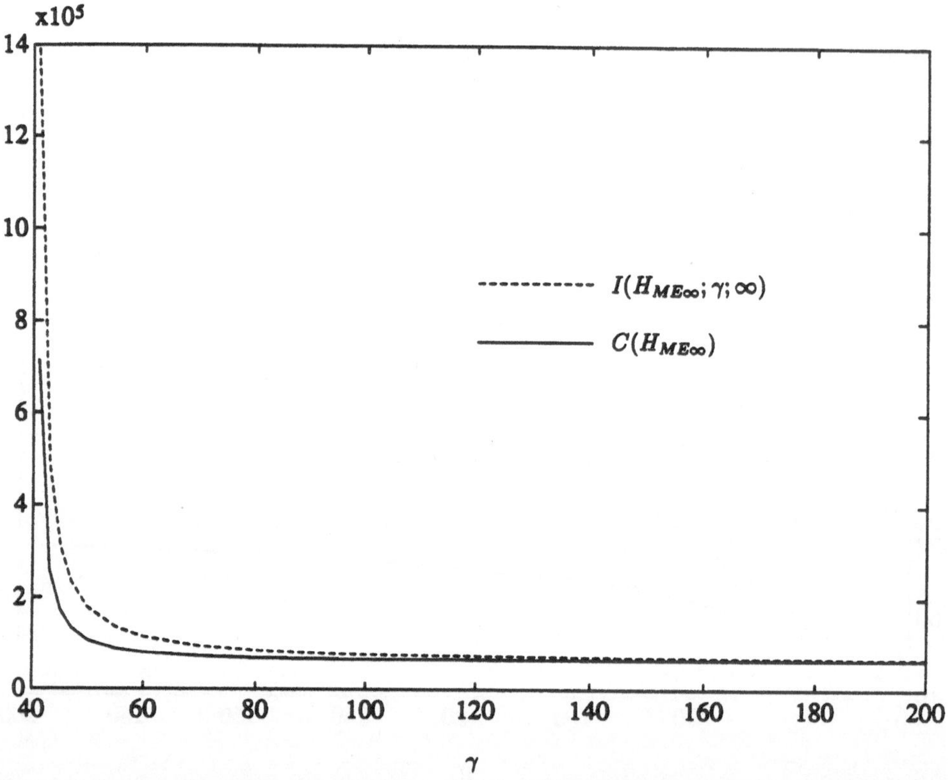

Figure 7.4: $C(H_{ME\infty})$ and its upper bound $I(H_{ME\infty};\gamma;\infty)$, against γ

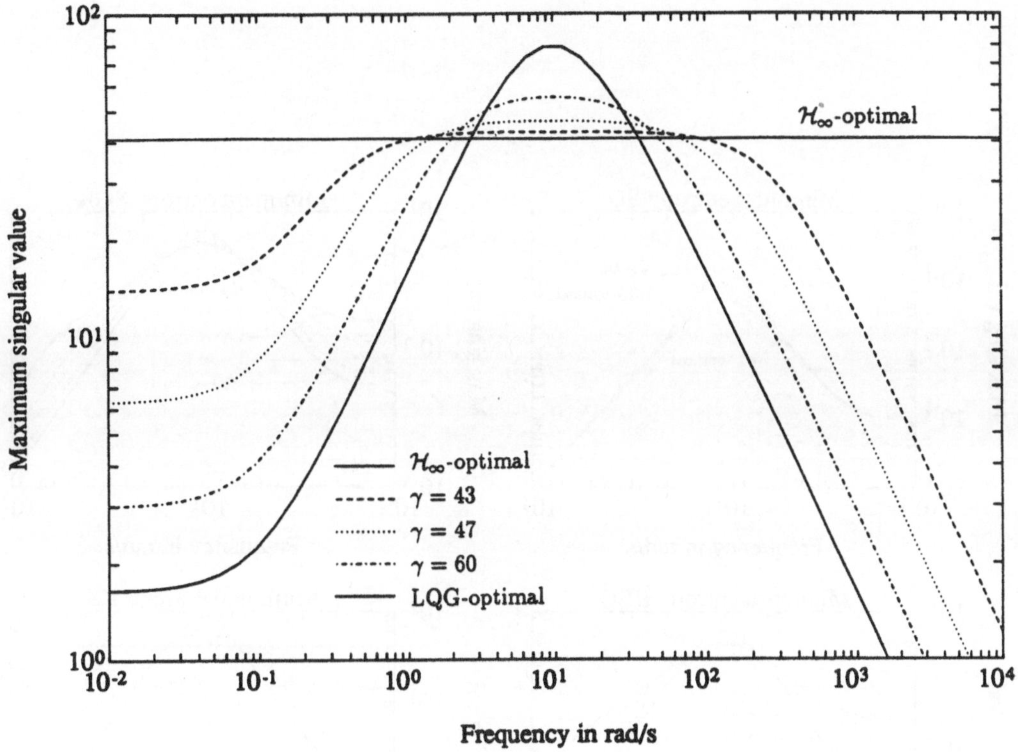

Figure 7.5: $H_{ME\infty}$ against frequency, for various γ

rise of the peak around 10 rad/s. Recalling from Remark 7.3.1 that the four transfer functions SG, SGK, KSG and KS which make up the closed-loop H all have robustness interpretations, it is of interest to plot these too—see Figure 7.6—again for the five representative values of γ. From these, Remark 7.3.1 shows us how to obtain a frequency-wise description of the maximum allowable uncertainty, for each of the uncertainties Δ_{ij} pictured in Figure 7.2. The maximum allowable uncertainties Δ_{ij} are plotted against frequency in Figure 7.7. Note that all of the maximum allowable uncertainties are larger at higher frequencies—robustness to high frequency uncertainty is good. This is what we would need in practice. Only in a narrow band of frequencies around 10 rad/s does the $\mathcal{H}_\infty$-optimal controller offer greater robustness than the minimum entropy/$\mathcal{H}_\infty$ controllers.

We can conclude that the (minimum entropy) Normalized $\mathcal{H}_\infty$ Controller with $\gamma = 60 \approx 1.5\gamma_o$ gives us nearly LQG-optimal performance, with robustness at least as good as the $\mathcal{H}_\infty$-optimal case except over a small band of frequencies around 10 rad/s. Examine Figures 7.3-7.7 for $\gamma = 60$ and compare with the $\mathcal{H}_\infty$-optimal and LQG-optimal cases.

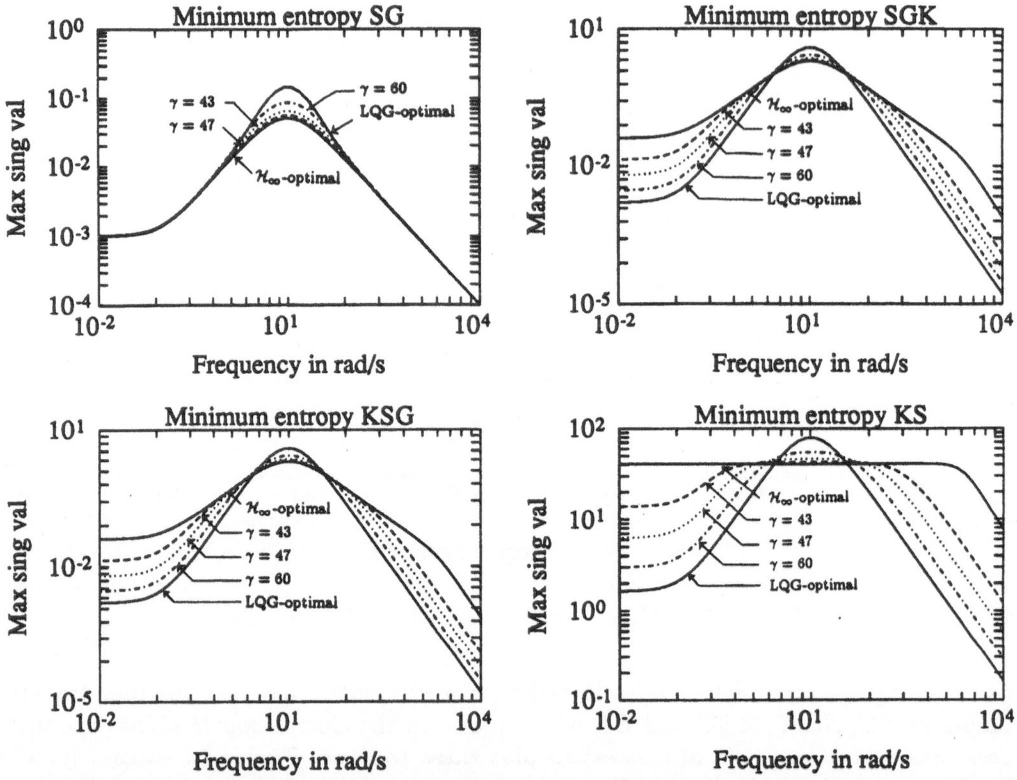

Figure 7.6: Components of $H_{ME\infty}$ against frequency, for various γ

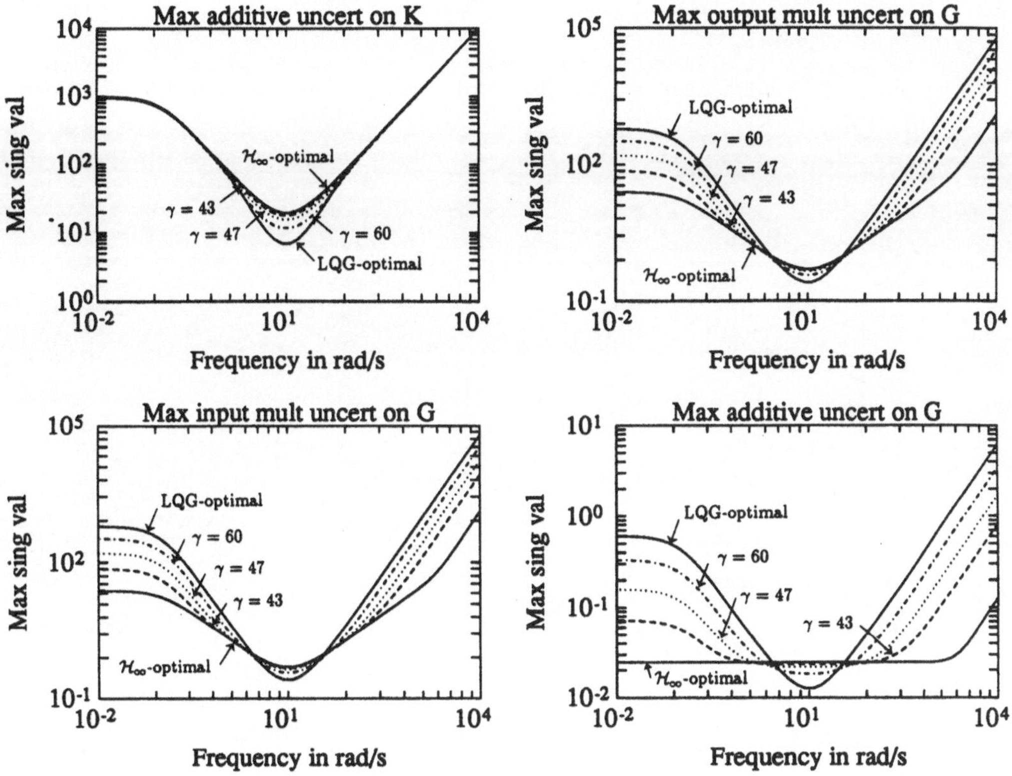

Figure 7.7: Maximum allowable uncertainties Δ_{ij} against frequency, for various γ

Chapter 8

$\mathcal{H}_\infty$-Characteristic Values

8.1 Introduction

The now well-established method of balancing the state-space description of a linear, asymptotically stable system was initiated in [44]. By representing a system in balanced coordinates, where the observability and controllability Gramians are equal and diagonal, the contribution of each state to the input-output map of the system is exposed. The Hankel singular values of the system, which are just the diagonal elements σ_i of the balanced Gramian, form a set of system input-output invariants. Each σ_i explicitly quantifies the relevance of its associated state, in that states corresponding to small σ_i are only weakly observable and weakly controllable. Deleting states corresponding to 'small' σ_i has a 'small' effect on the input-output map. Such 'balanced truncation' generically leads to a stable reduced-order model and the truncation error may be bounded in terms of the neglected σ_i (see [23, 26] and Lemma 8.4.4).

Open-loop balancing, however, has the restriction that the original system must be asymptotically stable. Furthermore, by its very nature, model reduction using standard balanced truncation takes no account of any closed-loop control considerations. These restrictions were lifted in [35], using a *closed*-loop balancing approach. The open-loop system (which may be unstable) is first compensated with the associated Normalized LQG Controller (Chapter 7). Two algebraic Riccati equations are needed—one for filtering and one for control. Balancing the solutions to these two Riccati equations, so that they are equal and diagonal, exposes the difficulty of filtering and controlling each state. To be exact, the diagonal elements μ_i of the solution to the 'LQG-balanced' Riccati equations form a set of closed-loop input-output invariants, known as the LQG-characteristic values. States corresponding to small μ_i are easy to filter and control; these states may be discarded with only a small effect on the closed-loop input-output map. For a full discussion of the properties of the μ_i, we refer the reader to [35]. An alternative (but by [46] equivalent) approach to closed-loop balancing and model reduction, based on coprime factorization, may be found in [41]; this will be discussed later.

Given the results of this monograph, and in particular given the results of Chapter 7, it is natural to ask whether an '$\mathcal{H}_\infty$-balancing' method is possible in the spirit of LQG-balancing. It is the purpose of this chapter to show that this is indeed the case. We will use the Normalized $\mathcal{H}_\infty$ Problem of Chapter 7 to provide the natural generalization of LQG-balancing to the minimum entropy/$\mathcal{H}_\infty$ case.

A different approach to obtaining reduced-order LQG or $\mathcal{H}_\infty$ controllers may be found in [33] and [9], respectively. There a fixed (reduced-order) structure for the controller is assumed and optimization theory is applied to derive four *coupled* matrix equations which the controller state-space matrices must satisfy (two Riccati equations and two Lyapunov equations in the LQG case, four Riccati equations in the $\mathcal{H}_\infty$ case). Our approach, in contrast, whilst not known to be optimal, only requires the pair of *decoupled* Riccati equations needed for the full-order Normalized $\mathcal{H}_\infty$ Controller.

The layout of the chapter is as follows. Section 8.2 summarises the LQG-balancing method of [35]. With Chapter 7 as background to the Normalized LQG Problem, we state the LQG-balancing method formally, and define the LQG-characteristic values, μ_i. In Section 8.3 the Normalized $\mathcal{H}_\infty$ Problem of Chapter 7 is used to introduce the new concept of $\mathcal{H}_\infty$-balancing, and the $\mathcal{H}_\infty$-characteristic values, ν_i, are defined. Some important properties of the ν_i are given. A new model reduction scheme—$\mathcal{H}_\infty$-balanced truncation—is presented and analysed in detail in Section 8.4: the ν_i play a major role.

8.2 LQG-Balancing and LQG-Characteristic Values

Consider the Normalized LQG Problem of Chapter 7 for an n-state minimal system $G = (A, B, C)$. The n-state Normalized LQG Controller solving this problem was given in Proposition 7.2.1—we need to solve the Control Algebraic Riccati equation (the CARE: equation (7.3))

$$A^T X_2 + X_2 A - X_2 BB^T X_2 + C^T C = 0 \qquad CARE$$

for the unique stabilizing solution $X_2 = X_2^T$, and we need to solve the Filter Algebraic Riccati equation (the FARE: equation (7.4))

$$AY_2 + Y_2 A^T - Y_2 C^T C Y_2 + BB^T = 0 \qquad FARE$$

for the unique stabilizing solution $Y_2 = Y_2^T$. Since $G = (A, B, C)$ is minimal, X_2 and Y_2 are positive definite (see Remark 7.2.2). Under a nonsingular state transformation S, $(A, B, C) \overset{S}{\mapsto} (SAS^{-1}, SB, CS^{-1}) =: (\tilde{A}, \tilde{B}, \tilde{C})$, so $X_2 \overset{S}{\mapsto} S^{-T} X_2 S^{-1} =: \tilde{X}_2$ and $Y_2 \overset{S}{\mapsto} SY_2S^T =: \tilde{Y}_2$. Since $\tilde{X}_2\tilde{Y}_2 = S^{-T}X_2Y_2S^T$, it follows that X_2Y_2 and $\tilde{X}_2\tilde{Y}_2$ are similar matrices, and therefore the eigenvalues of X_2Y_2 are similarity invariants. These invariants are the squares of the *LQG-characteristic values* of the system G, as defined in [35]; a formal definition and some basic properties are given next.

Proposition 8.2.1 (LQG-balancing and LQG-characteristic values [35])
Let the system $G = (A, B, C)$ be minimal with n-states and let X_2 and Y_2 be the unique positive definite stabilizing solutions of the CARE and FARE respectively. Then the eigenvalues of X_2Y_2 are real, strictly positive similarity invariants, as are their positive square roots which are called the LQG-characteristic values *of G. Let $\mu_1^2 \geq \mu_2^2 \geq \cdots \geq \mu_n^2 > 0$ denote the n eigenvalues of X_2Y_2 arranged in decreasing order, then there exists a similarity transformation which transforms both X_2 and Y_2 to the form $M := \text{diag}(\mu_1, \mu_2, \ldots, \mu_n)$. The system is then said to be in* LQG-balanced coordinates, *and M is the diagonal matrix of LQG-characteristic values of G.*

Remark 8.2.2 (Uniqueness of LQG-balanced realizations)
In [35], it was shown that an LQG-balanced realization of a system is unique to within block diagonal similarity transformations. The block diagonal elements are: ± 1 for each distinct μ_i, and any l by l orthogonal matrix for each μ_i with multiplicity l.

Ordering of the blocks is done compatibly with the ordering of the μ_i. In the generic case where the μ_i are all distinct, the LQG-balanced realization is therefore unique to within similarity transformation by a sign matrix.

As argued in [35], small μ_i correspond to states which are easy to filter and control. This motivates the following model reduction schemes.

Procedure 8.2.3 (Reduced-order plant by LQG-balanced truncation [35]) *Let $G = (A, B, C)$ be minimal with n-states and in LQG-balanced coordinates with LQG-characteristic values $\mu_1 \geq \mu_2 \geq \cdots \geq \mu_n > 0$. That is, $M = \text{diag}(\mu_1, \mu_2, \ldots, \mu_n)$ is the stabilizing solution of the CARE and FARE associated with $G = (A, B, C)$. Pick $k < n$ such that $\mu_k > \mu_{k+1}$ and partition M accordingly into*

$$M = \begin{bmatrix} M_1 & 0 \\ 0 & M_2 \end{bmatrix}$$

where $M_1 = \text{diag}(\mu_1, \ldots, \mu_k)$ and $M_2 = \text{diag}(\mu_{k+1}, \ldots, \mu_n)$. Partition A, B and C conformably with the partitioning of M:

$$A = \begin{bmatrix} A_{11} & A_{12} \\ A_{21} & A_{22} \end{bmatrix}, \qquad B = \begin{bmatrix} B_1 \\ B_2 \end{bmatrix} \quad \text{and} \quad C = \begin{bmatrix} C_1 & C_2 \end{bmatrix}. \tag{8.1}$$

A k-state reduced-order plant is then $G_r = (A_{11}, B_1, C_1)$.

Procedure 8.2.4 (Reduced-order control by LQG-balanced truncation [35]) *Let $G = (A, B, C)$ be minimal with n-states and in LQG-balanced coordinates with LQG-characteristic values $\mu_1 \geq \mu_2 \geq \cdots \geq \mu_n > 0$. That is, $M = \text{diag}(\mu_1, \mu_2, \ldots, \mu_n)$ is the stabilizing solution of the CARE and FARE associated with $G = (A, B, C)$. Pick $k < n$ such that $\mu_k > \mu_{k+1}$ and partition M accordingly into*

$$M = \begin{bmatrix} M_1 & 0 \\ 0 & M_2 \end{bmatrix}$$

where $M_1 = \text{diag}(\mu_1, \ldots, \mu_k)$ and $M_2 = \text{diag}(\mu_{k+1}, \ldots, \mu_n)$. Let $K_{LQG} = (\hat{A}, \hat{B}, \hat{C})$ be the Normalized LQG Controller for the plant $G = (A, B, C)$ (as given in Proposition 7.2.1). Partition $\hat{A}$, $\hat{B}$ and $\hat{C}$ conformably with the partitioning of M:

$$\hat{A} = \begin{bmatrix} \hat{A}_{11} & \hat{A}_{12} \\ \hat{A}_{21} & \hat{A}_{22} \end{bmatrix}, \qquad \hat{B} = \begin{bmatrix} \hat{B}_1 \\ \hat{B}_2 \end{bmatrix} \quad \text{and} \quad \hat{C} = \begin{bmatrix} \hat{C}_1 & \hat{C}_2 \end{bmatrix}.$$

A k-state reduced-order controller is then $K_r = (\hat{A}_{11}, \hat{B}_1, \hat{C}_1)$.

Remark 8.2.5 The reduced-order controller K_r is the full-order Normalized LQG Controller for the reduced-order plant G_r. This is an immediate consequence of the easily seen fact that M_1 is the stabilizing solution to the CARE and FARE for $G_r = (A_{11}, B_1, C_1)$.

In the remainder of this chapter, we will be concerned with various ways of obtaining reduced-order plants and controllers. To keep notation simple, we will use G_r to denote a reduced-order plant and will state explicitly which model reduction method was used to obtain it. Similarly for a reduced-order controller K_r.

Of course, it is important to know what happens to the stability and performance of the closed-loop when a reduced-order Normalized LQG Controller is connected to the full-order system (A, B, C). Sufficient conditions are derived in [35]; they confirm the intuitive notion that the reduced-order controller is likely to perform well if the discarded μ_i are sufficiently small.

8.3 $\mathcal{H}_\infty$-Balancing and $\mathcal{H}_\infty$-Characteristic Values

In this section, we show how all the results of the previous section can be generalized to the minimum entropy/$\mathcal{H}_\infty$ case. Consider the Normalized $\mathcal{H}_\infty$ Problem of Chapter 7 for an n-state minimal system $G = (A, B, C)$. The n-state Normalized $\mathcal{H}_\infty$ Controller solving this problem was given in Proposition 7.3.3—we need to solve the $\mathcal{H}_\infty$ Control Algebraic Riccati equation (the HCARE: equation (7.9))

$$A^T X_\infty + X_\infty A - (1-\gamma^{-2}) X_\infty B B^T X_\infty + C^T C = 0 \qquad HCARE$$

for the unique stabilizing solution $X_\infty = X_\infty^T$, and we need to solve the $\mathcal{H}_\infty$ Filter Algebraic Riccati equation (the HFARE: equation (7.10))

$$A Y_\infty + Y_\infty A^T - (1-\gamma^{-2}) Y_\infty C^T C Y_\infty + B B^T = 0 \qquad HFARE$$

for the unique stabilizing solution $Y_\infty = Y_\infty^T$. Since $G = (A, B, C)$ is minimal, X_∞ and Y_∞ are positive definite (see Remark 7.3.4). Just as in the LQG case, under a nonsingular state transformation S, $X_\infty Y_\infty \overset{S}{\mapsto} S^{-T} X_\infty Y_\infty S^T$, so the eigenvalues of $X_\infty Y_\infty$ are similarity invariants, the positive square roots of which we define to be the *$\mathcal{H}_\infty$-characteristic values*. All the arguments of Section 8.2 carry across and we obtain the following minimum entropy/$\mathcal{H}_\infty$ generalization of Proposition 8.2.1. Note that since $\gamma > \gamma_o$ by assumption, Proposition 7.3.3 gives $\rho(X_\infty Y_\infty) < \gamma^2$ so each of the $\mathcal{H}_\infty$-characteristic values is strictly less than γ.

Proposition 8.3.1 ($\mathcal{H}_\infty$-balancing and $\mathcal{H}_\infty$-characteristic values)
Let the system $G = (A, B, C)$ be minimal with n-states, let $\gamma > \gamma_o$, and let X_∞ and Y_∞ be the unique positive definite stabilizing solutions of the HCARE and HFARE respectively. Then the eigenvalues of $X_\infty Y_\infty$ are real, strictly positive similarity invariants, as are their positive square roots which are called the $\mathcal{H}_\infty$-characteristic values *of G. Let $\nu_1^2 \geq \nu_2^2 \geq \cdots \geq \nu_n^2 > 0$ denote the n eigenvalues of $X_\infty Y_\infty$ arranged in decreasing order, then $\gamma > \nu_i$ and there exists a similarity transformation which transforms both X_∞ and Y_∞ to the form $N := \mathrm{diag}(\nu_1, \nu_2, \ldots, \nu_n)$. The system is then said to be in* $\mathcal{H}_\infty$-balanced coordinates, *and N is the diagonal matrix of $\mathcal{H}_\infty$-characteristic values of G.*

Remark 8.3.2 (Uniqueness of $\mathcal{H}_\infty$-balanced realizations) It is easy to see that an $\mathcal{H}_\infty$-balanced realization of a system is unique to within similarity transformations of an analogous type to those described in Remark 8.2.2 for LQG-balanced realizations. In particular, we note that in the generic case where the ν_i are all distinct, the $\mathcal{H}_\infty$-balanced realization is unique to within a similarity transformation by a sign matrix.

Note that the $\mathcal{H}_\infty$-characteristic values are functions of γ. Strictly speaking, we should write $\nu_i(\gamma)$ and $N(\gamma)$. In the interests of notational simplicity, we will write ν_i and N.

The following proposition states some interesting properties of the ν_i.

Proposition 8.3.3 *For the ν_i as defined in Proposition 8.3.1 we have*

(i) $\nu_i \geq \mu_i$.

(ii) *Each ν_i is a monotonically decreasing function of γ.*

(iii) *If the ν_i are distinct then $d\nu_i/d\gamma \leq 0$.*

(iv) *Each ν_i is a continuous function of γ.*

(v) $\lim_{\gamma\to\infty}\{\nu_i\} = \mu_i$.

Proof Note firstly that X_∞ and Y_∞ are positive definite, hence non-singular, because (A, B, C) is minimal by assumption (Remark 7.3.4).

Part (i) Apply the main result of [62] to $X_\infty^{-1}(\text{HCARE})X_\infty^{-1}$ and $X_2^{-1}(\text{CARE})X_2^{-1}$, and then to $Y_\infty^{-1}(\text{HFARE})Y_\infty^{-1}$ and $Y_2^{-1}(\text{FARE})Y_2^{-1}$. It follows that

$$X_\infty \geq X_2$$

and

$$Y_\infty \geq Y_2.$$

Hence [32, Observation 7.7.2] we have

$$X_\infty^{1/2} Y_\infty X_\infty^{1/2} \geq X_\infty^{1/2} Y_2 X_\infty^{1/2}$$

and

$$Y_2^{1/2} X_\infty Y_2^{1/2} \geq Y_2^{1/2} X_2 Y_2^{1/2},$$

which implies [32, Corollary 7.7.4(c)] that

$$\lambda_i\{X_\infty Y_\infty\} \geq \lambda_i\{X_\infty Y_2\} \geq \lambda_i\{X_2 Y_2\}.$$

But, by definition, $\nu_i^2 = \lambda_i\{X_\infty Y_\infty\}$ and $\mu_i^2 = \lambda_i\{X_2 Y_2\}$, and the result follows.

Part (ii) Let $\gamma_2 \geq \gamma_1 > \gamma_o$. Apply the main result of [62] again, this time to $X_\infty^{-1}(\text{HCARE})X_\infty^{-1}$ with $\gamma = \gamma_1$ and $\gamma = \gamma_2$, and then to $Y_\infty^{-1}(\text{HFARE})Y_\infty^{-1}$ with $\gamma = \gamma_1$ and $\gamma = \gamma_2$. It follows that

$$X_\infty(\gamma_2) \leq X_\infty(\gamma_1)$$

and

$$Y_\infty(\gamma_2) \leq Y_\infty(\gamma_1),$$

(with an obvious notation). Using a similar argument to that in the proof of Part (i), this implies that $\nu_i(\gamma_2) \leq \nu_i(\gamma_1)$, and the result is proved. In fact, from [14, 50], X_∞ and Y_∞ are differentiable functions of γ, so we actually have

$$\frac{dX_\infty}{d\gamma} \leq 0$$

and

$$\frac{dY_\infty}{d\gamma} \leq 0.$$

Part (iii) From the proof of Part (ii), X_∞ and Y_∞ are differentiable functions of γ. Hence so is $X_\infty Y_\infty$. By definition, $\nu_i^2 = \lambda_i\{X_\infty Y_\infty\}$ and, by assumption, the ν_i are distinct. Hence each ν_i is a differentiable function of γ. But, from Part (ii), each ν_i is a monotonically decreasing function of γ. Therefore, for each ν_i, there holds $d\nu_i/d\gamma \leq 0$, as claimed.

Part (iv) From the proof of Part (iii), $X_\infty Y_\infty$ is a differentiable function of γ. Hence it is a continuous function of γ. The result then follows from the well-known fact that the eigenvalues of a matrix are continuous functions of the elements of the matrix.

Part (v) From Parts (i), (ii) and (iv), each ν_i is a continuous, monotonically decreasing function of γ, bounded below by μ_i. Hence $\lim_{\gamma\to\infty}\{\nu_i\}$ exists. It is easy to see that as $\gamma \to \infty$, the HCARE tends to the CARE, and the HFARE tends to the FARE. So,

$$\lim_{\gamma\to\infty}\{X_\infty\} = X_2$$

and

$$\lim_{\gamma\to\infty}\{Y_\infty\} = Y_2.$$

Recalling that $\nu_i^2 = \lambda_i\{X_\infty Y_\infty\}$ and $\mu_i^2 = \lambda_i\{X_2 Y_2\}$, the result follows. □

Remark 8.3.4 In [47], it was shown that the LQG-balancing approach is particularly applicable to symmetric passive systems, such as large space structures with colocated rate sensors and actuators. We would expect this to be the case for the $\mathcal{H}_\infty$-balancing approach, too. However, as pointed out in [47, p83], for systems with no symmetry or passivity properties, a non-normalized approach may be more suitable. One possible way round this would be to apply the normalized approach not to the actual plant, but to a 'shaped' plant which has been (open-loop) pre- and post-compensated. The success of such 'loop-shaping' ideas has been demonstrated in [40], using controller synthesis via the robust stabilization of the normalized coprime factors of a plant.

Remark 8.3.5 In [35], there is a parametrization, in the case of distinct μ_i, of single-input single-output systems in terms of the μ_i and the elements of the B-vector in the realization of G. Using properties of this parametrization, necessary and sufficient conditions were derived for dissipativeness of both the open-loop and closed-loop, with either the full-order or reduced-order controller. Note that necessary and sufficient conditions for dissipativeness provide sufficient conditions for stability. These results all carry across to the minimum entropy/$\mathcal{H}_\infty$ case in a straightforward way, by using the ν_i and making the appropriate changes.

8.4 Model Reduction by $\mathcal{H}_\infty$-Balanced Truncation

Let us now turn to the model reduction aspects of the Normalized $\mathcal{H}_\infty$ Problem. Controller or plant order reduction may be carried out using *$\mathcal{H}_\infty$-balanced truncation*, an analogous approach to the LQG-balanced truncation of Procedure 8.2.4 and Procedure 8.2.3: write the plant and its Normalized $\mathcal{H}_\infty$ Controller in $\mathcal{H}_\infty$-balanced coordinates, and discard those states in either the plant or controller associated with 'small' ν_i. We now state this new *$\mathcal{H}_\infty$-balanced truncation* method formally.

8.4.1 $\mathcal{H}_\infty$-Balanced Truncation

Procedure 8.4.1 (Reduced-order plant by $\mathcal{H}_\infty$-balanced truncation)
Let $G = (A, B, C)$ be minimal with n-states and in $\mathcal{H}_\infty$-balanced coordinates for given $\gamma > \gamma_o$, with $\mathcal{H}_\infty$-characteristic values $\gamma > \nu_1 \geq \nu_2 \geq \cdots \geq \nu_n > 0$. That is, $N = \operatorname{diag}(\nu_1, \nu_2, \ldots, \nu_n)$ is the stabilizing solution of the HCARE and HFARE associated with $G = (A, B, C)$ and γ. Pick $k < n$ such that $\nu_k > \nu_{k+1}$ and partition N accordingly into

$$N = \begin{bmatrix} N_1 & 0 \\ 0 & N_2 \end{bmatrix}$$

where $N_1 = \operatorname{diag}(\nu_1, \ldots, \nu_k)$ and $N_2 = \operatorname{diag}(\nu_{k+1}, \ldots, \nu_n)$. Partition A, B and C conformably with the partitioning of N:

$$A = \begin{bmatrix} A_{11} & A_{12} \\ A_{21} & A_{22} \end{bmatrix}, \quad B = \begin{bmatrix} B_1 \\ B_2 \end{bmatrix} \quad \text{and} \quad C = \begin{bmatrix} C_1 & C_2 \end{bmatrix}.$$

A k-state reduced-order plant is then $G_r = (A_{11}, B_1, C_1)$.

Procedure 8.4.2 (Reduced-order control by $\mathcal{H}_\infty$-balanced truncation)
Let $G = (A, B, C)$ be minimal with n-states and in $\mathcal{H}_\infty$-balanced coordinates for given $\gamma > \gamma_o$, with $\mathcal{H}_\infty$-characteristic values $\gamma > \nu_1 \geq \nu_2 \geq \cdots \geq \nu_n > 0$. That is, $N = \operatorname{diag}(\nu_1, \nu_2, \ldots, \nu_n)$ is the stabilizing solution of the HCARE and HFARE associated with $G = (A, B, C)$ and γ. Pick $k < n$ such that $\nu_k > \nu_{k+1}$ and partition N accordingly into

$$N = \begin{bmatrix} N_1 & 0 \\ 0 & N_2 \end{bmatrix}$$

where $N_1 = \mathrm{diag}(\nu_1, \ldots, \nu_k)$ *and* $N_2 = \mathrm{diag}(\nu_{k+1}, \ldots, \nu_n)$. *Let* $K_{ME\infty} = (\hat{A}, \hat{B}, \hat{C})$ *be the Normalized* $\mathcal{H}_\infty$ *Controller for the plant* $G = (A, B, C)$ *and* γ *(as given in Proposition 7.3.3). Partition* $\hat{A}$, $\hat{B}$ *and* $\hat{C}$ *conformably with the partitioning of* N*:*

$$\hat{A} = \begin{bmatrix} \hat{A}_{11} & \hat{A}_{12} \\ \hat{A}_{21} & \hat{A}_{22} \end{bmatrix}, \qquad \hat{B} = \begin{bmatrix} \hat{B}_1 \\ \hat{B}_2 \end{bmatrix} \quad \text{and} \quad \hat{C} = \begin{bmatrix} \hat{C}_1 & \hat{C}_2 \end{bmatrix}.$$

A k*-state reduced-order controller is then* $K_r = (\hat{A}_{11}, \hat{B}_1, \hat{C}_1)$.

Remark 8.4.3 The reduced-order controller K_r is the full-order Normalized $\mathcal{H}_\infty$ Controller for the reduced-order plant G_r. This is an immediate consequence of the easily seen fact that N_1 is the stabilizing solution to the HCARE and HFARE for $G_r = (A_{11}, B_1, C_1)$.

8.4.2 Relations to Balanced Truncation

Here we relate $\mathcal{H}_\infty$-balancing to ordinary balancing. This sets the scene for later sections, where a precise connection with balanced truncation of a scaled factorization of the plant is derived.

Suppose we carry out ordinary balanced truncation [44] on the open-loop plant G. Assume $G = (A, B, C)$ is asymptotically stable and minimal and in balanced coordinates. That is, the controllability Gramian and the observability Gramian are both equal to the *balanced Gramian* Σ, where $\Sigma := \mathrm{diag}(\sigma_1, \sigma_2, \ldots, \sigma_n) > 0$ solves

$$\begin{aligned} A\Sigma + \Sigma A^T + BB^T &= 0 \\ A^T\Sigma + \Sigma A + C^T C &= 0. \end{aligned}$$

Any asymptotically stable and minimal system may be balanced by a suitable similarity transform [44]. The matrix Σ is just the diagonal matrix of *Hankel singular values*; by convention we order them $\sigma_1 \geq \sigma_2 \geq \cdots \geq \sigma_n > 0$. Choose $k < n$ and partition

$$\Sigma = \begin{bmatrix} \Sigma_1 & 0 \\ 0 & \Sigma_2 \end{bmatrix}$$

where $\Sigma_1 = \mathrm{diag}(\sigma_1, \ldots, \sigma_k)$ and $\Sigma_2 = \mathrm{diag}(\sigma_{k+1} \ldots, \sigma_n)$. Partition the plant state-space matrices conformably in the obvious way (as in (8.1)). Then a reduced-order plant is $G_r = (A_{11}, B_1, C_1)$, obtained by deleting the states associated with Σ_2.

Lemma 8.4.4 *Let* G *be asymptotically stable and minimal with* n*-states. Let* G *be in balanced coordinates with balanced Gramian* $\Sigma = \mathrm{diag}(\sigma_1, \sigma_2, \ldots, \sigma_n)$ *and Hankel singular values* $\sigma_1 \geq \sigma_2 \geq \cdots \geq \sigma_n > 0$. *Let* $k < n$ *and partition* $\Sigma = \mathrm{diag}(\Sigma_1, \Sigma_2)$ *where* $\Sigma_1 = \mathrm{diag}(\sigma_1, \ldots, \sigma_k)$ *and* $\Sigma_2 = \mathrm{diag}(\sigma_{k+1} \ldots, \sigma_n)$. *Let* G_r *be the* k*-state reduced-order model obtained by truncating the balanced realization of* G *to* k*-states. Then*

(i) [49] *G_r is in balanced coordinates with balanced Gramian Σ_1.*

(ii) [49] *If $\sigma_k > \sigma_{k+1}$ then G_r is asymptotically stable and minimal.*

(iii) [26, 23] $\|G - G_r\|_\infty \leq 2\,\mathrm{trace}[\Sigma_2]$.

We can now relate the $\mathcal{H}_\infty$-characteristic values of G to the Hankel singular values of G.

Proposition 8.4.5 *Suppose $G = (A, B, C)$ is asymptotically stable and minimal, with Hankel singular values $\sigma_1 \geq \sigma_2 \geq \cdots \geq \sigma_n > 0$. Let $\gamma > \gamma_o$ and let G have $\mathcal{H}_\infty$-characteristic values $\gamma > \nu_1 \geq \nu_2 \geq \cdots \geq \nu_n > 0$. Then*

(i) *If $\gamma > 1$ then $\sigma_i \geq \nu_i$.*

(ii) *If $\gamma = 1$ then $\sigma_i = \nu_i$.*

(iii) *If $\gamma < 1$ then $\sigma_i \leq \nu_i$.*

Proof In this proof, (A, B, C) is not necessarily a balanced realization of any type. The controllability Gramian $P = P^T > 0$ and observability Gramian $Q = Q^T > 0$ of $G = (A, B, C)$ are given by

$$AP + PA^T + BB^T = 0 \tag{8.2}$$
$$A^TQ + QA + C^TC = 0, \tag{8.3}$$

and then

$$\sigma_i^2 = \lambda_i\{QP\}.$$

Subtract the HFARE from (8.2), and subtract the HCARE from (8.3):

$$A(P - Y_\infty) + (P - Y_\infty)A^T + (1 - \gamma^{-2})Y_\infty C^TCY_\infty = 0 \tag{8.4}$$
$$A^T(Q - X_\infty) + (Q - X_\infty)A + (1 - \gamma^{-2})X_\infty BB^TX_\infty = 0. \tag{8.5}$$

Part (i) By assumption, A is asymptotically stable and $1 - \gamma^{-2} > 0$. A standard result on Lyapunov equations (see [26, Theorem 3.3(7)]) applied to (8.4) and (8.5) then gives $P \geq Y_\infty$ and $Q \geq X_\infty$. Hence, by the argument in the proof of Proposition 8.3.3(i), $\lambda_i\{QP\} \geq \lambda_i\{X_\infty Y_\infty\}$. Recalling $\sigma_i^2 = \lambda_i\{QP\}$ and $\nu_i^2 = \lambda_i\{X_\infty Y_\infty\}$, the result follows.

Part (ii) Put $\gamma = 1$ in (8.4) and (8.5):

$$A(P - Y_\infty) + (P - Y_\infty)A^T = 0$$
$$A^T(Q - X_\infty) + (Q - X_\infty)A = 0.$$

Since A is asymptotically stable by assumption, it follows from [48] that $P = Y_\infty$ and $Q = X_\infty$. Hence the result.

Part (iii) The same argument as used to prove Part (i) may be applied, except now $1-\gamma^{-2} < 0$. It follows that $P \leq Y_\infty$ and $Q \leq X_\infty$, which implies the result. □

The following corollary is immediate on combining Parts (ii) and (iii) of the proposition. (Minimality of $G = (A, B, C)$ is relaxed to stabilizability and detectability in the corollary because this does not affect γ_o, σ_1 or ν_1.)

Corollary 8.4.6 *Suppose $G = (A, B, C)$ is asymptotically stable, stabilizable and detectable, with largest Hankel singular value $\sigma_1 > 0$. Then $\gamma_o < 1$ only if $\sigma_1 < 1$.*

We recognise σ_1 as the *Hankel norm* [26] of the system G.

Now suppose we carry out balanced truncation on the normalized coprime factors of the plant (see [41] and also [40, Chapter 5]). To be precise, let G have a normalized left-coprime factorization:

$$G = \tilde{M}^{-1}\tilde{N},$$

where $\tilde{M}$, $\tilde{N}$ are asymptotically stable and left-coprime, $\tilde{M}^{-1}$ exists, and

$$\tilde{N}\tilde{N}^* + \tilde{M}\tilde{M}^* = I.$$

Let the Hankel singular values of $[\tilde{N} \;\; \tilde{M}]$ be $\tilde{\sigma}_1 \geq \tilde{\sigma}_2 \geq \cdots \geq \tilde{\sigma}_n > 0$ and define $\tilde{\Sigma} = \mathrm{diag}(\tilde{\sigma}_1, \tilde{\sigma}_2, \ldots, \tilde{\sigma}_n)$. Since the factorization is normalized, we have [41] that $I > \tilde{\Sigma}$, so $1 > \tilde{\sigma}_i$. Pick $k < n$ such that $\tilde{\sigma}_k > \tilde{\sigma}_{k+1}$. We can then perform balanced truncation in the usual way (i.e., as described above) on $[\tilde{N} \;\; \tilde{M}]$ to obtain a reduced-order representation $[\tilde{N}_r \;\; \tilde{M}_r]$. It was proved in [41] that $G_r = \tilde{M}_r^{-1}\tilde{N}_r$ is in fact a normalized left-coprime factorization of the reduced-order plant G_r. The error bound in Lemma 8.4.4(iii) applies:

$$\|[\tilde{N} - \tilde{N}_r \;\; \tilde{M} - \tilde{M}_r]\|_\infty \leq 2 \;\mathrm{trace}[\tilde{\Sigma}_2],$$

this time in terms of $\tilde{\Sigma}_2$, the diagonal matrix of the neglected Hankel singular values of $[\tilde{N} \;\; \tilde{M}]$.

It is well-known [42] that the solution to a suitable Linear Quadratic Regulator problem leads to the state-space realization of the normalized left-coprime factors. So it is to be expected that the Hankel singular values of $[\tilde{N} \;\; \tilde{M}]$ are closely related to the LQG-characteristic values.

Lemma 8.4.7 ([40, p80]) *Let $1 > \tilde{\sigma}_1 \geq \tilde{\sigma}_2 \geq \cdots \geq \tilde{\sigma}_n > 0$ be the Hankel singular values of $[\tilde{N} \;\; \tilde{M}]$. Then*

$$\tilde{\sigma}_i^2 = \frac{\mu_i^2}{1 + \mu_i^2}.$$

From Proposition 8.3.3(i), $\nu_i \geq \mu_i$. The following result, which relates the $\mathcal{H}_\infty$-characteristic values of G to the Hankel singular values of $[\tilde{N} \;\; \tilde{M}]$, is immediate from this fact together with Lemma 8.4.7.

Proposition 8.4.8 *Definitions as above. Then*

$$\tilde{\sigma}_i^2 \leq \frac{\nu_i^2}{1 + \nu_i^2}.$$

The above is a link between $\mathcal{H}_\infty$-balancing and coprime factorization of the plant G. In the next section we shall see a much stronger link with coprime factorization of a scaled plant.

8.4.3 Coprime Factorization via the Normalized $\mathcal{H}_\infty$ Problem

Motivated by the relations of the previous section, we derive here a much more precise interpretation of the ν_i: we find that the Hankel singular values of the normalized left-coprime factors of a *scaled* plant can be written exactly in terms of the ν_i.

In Proposition 8.4.8, the $\mathcal{H}_\infty$-characteristic values were related to the Hankel singular values of the normalized left-coprime factors of the plant G. A key step was to use [42] to construct the normalized left-coprime factorization of G using one of the LQG algebraic Riccati equations of the Normalized LQG Problem. The key step in this section is to use [42] to construct the normalized left-coprime factorization of βG using one of the $\mathcal{H}_\infty$ algebraic Riccati equations of the Normalized $\mathcal{H}_\infty$ Problem. Here

$$\beta := \sqrt{1-\gamma^{-2}}$$

is a necessary plant scaling—we assume $\gamma > \max\{1, \gamma_o\}$ (refer to Lemma 7.3.6 and Corollary 8.4.6), so that $0 < \beta \leq 1$.

Lemma 8.4.9 *Let $G = (A, B, C)$ be minimal and let $\gamma > \max\{1, \gamma_o\}$. Let $\beta = (1-\gamma^{-2})^{1/2}$. Let Y_∞ be the unique positive definite stabilizing solution of the associated HFARE. Define $\bar{N}$ and $\bar{M}$ by*

$$[\beta\bar{N} \;\; \bar{M}] = \left[\begin{array}{c|c} \bar{A} & \bar{B} \\ \hline \bar{C} & \bar{D} \end{array}\right] := \left[\begin{array}{c|cc} A - \beta^2 Y_\infty C^T C & \beta B & -\beta^2 Y_\infty C^T \\ \hline C & 0 & I \end{array}\right]. \tag{8.6}$$

Then $\beta G = \bar{M}^{-1}\beta\bar{N}$ is a normalized left-coprime factorization of βG.

Proof Appendix A.5. □

Remark 8.4.10 From [46, 43], the realization (8.6) of $[\beta\bar{N} \;\; \bar{M}]$ is minimal if and only if the realization $G = (A, B, C)$ is minimal. Minimality of $G = (A, B, C)$ is a standing assumption in this chapter.

In order to work out the Hankel singular values of $[\beta\bar{N} \;\; \bar{M}]$ we need to calculate the controllability Gramian $\bar{P}$ of $[\beta\bar{N} \;\; \bar{M}]$ and the observability Gramian $\bar{Q}$ of $[\beta\bar{N} \;\; \bar{M}]$. Since the realization of $[\beta\bar{N} \;\; \bar{M}]$ given in (8.6) is asymptotically stable and minimal, both $\bar{P}$ and $\bar{Q}$ are positive definite. In fact, these Gramians can be written explicitly in terms of X_∞ (the unique positive definite stabilizing solution of the HCARE) and Y_∞ (the unique positive definite stabilizing solution of the HFARE).

Lemma 8.4.11 *Consider the state-space realization of $[\beta\bar{N} \;\; \bar{M}]$ given in Lemma 8.4.9. Its controllability Gramian is*

$$\bar{P} = \beta^2 Y_\infty,$$

and its observability Gramian is

$$\bar{Q} = (X_\infty^{-1} + \beta^2 Y_\infty)^{-1}.$$

Proof The controllability Gramian $\bar{P} = \bar{P}^T$ is the unique positive definite solution to the Lyapunov equation

$$\bar{P}(A - \beta^2 Y_\infty C^T C)^T + (A - \beta^2 Y_\infty C^T C)\bar{P} + \beta^2 BB^T + \beta^4 Y_\infty C^T C Y_\infty = 0.$$

It is easy to show, using the HFARE, that $\bar{P} = \beta^2 Y_\infty$ solves this equation.

The observability Gramian $\bar{Q} = \bar{Q}^T$ is the unique positive definite solution to the Lyapunov equation

$$\bar{Q}(A - \beta^2 Y_\infty C^T C) + (A - \beta^2 Y_\infty C^T C)^T \bar{Q} + C^T C = 0.$$

Pre- and post-multiply by $\bar{Q}^{-1}$ and substitute $\bar{Q}^{-1} = X_\infty^{-1} + \beta^2 Y_\infty$. It is then straightforward to use the HFARE and the HCARE to show that the equation is indeed solved by this choice of $\bar{Q}$. □

The Hankel singular values of $[\beta\bar{N} \;\; \bar{M}]$ are

$$\bar{\sigma}_i^2 = \lambda_i\{\bar{Q}\bar{P}\},$$

ordered $1 > \bar{\sigma}_1 \geq \bar{\sigma}_2 \geq \cdots \geq \bar{\sigma}_n > 0$. Lemma 8.4.11 immediately allows the $\bar{\sigma}_i$ to be expressed in terms of the $\mathcal{H}_\infty$-characteristic values ν_i.

Proposition 8.4.12 *Let $\beta G = \bar{M}^{-1}\beta\bar{N}$ be the normalized left-coprime factorization given in Lemma 8.4.9. Let $1 > \bar{\sigma}_1 \geq \bar{\sigma}_2 \geq \cdots \geq \bar{\sigma}_n > 0$ be the Hankel singular values of $[\beta\bar{N} \;\; \bar{M}]$. Let $\gamma > \nu_1 \geq \nu_2 \geq \cdots \geq \nu_n > 0$ be the $\mathcal{H}_\infty$-characteristic values of G, where $\gamma > \max\{1, \gamma_o\}$. Then*

$$\bar{\sigma}_i^2 = \frac{\beta^2 \nu_i^2}{1 + \beta^2 \nu_i^2}.$$

Proof We firstly recall the definitions $\bar{\sigma}_i^2 = \lambda_i\{\bar{Q}\bar{P}\}$ and $\nu_i^2 = \lambda_i\{X_\infty Y_\infty\}$. Then, using Lemma 8.4.11 to express $\bar{P}$ and $\bar{Q}$ in terms of X_∞ and Y_∞, we have

$$\begin{aligned}
\bar{\sigma}_i^2 &= \lambda_i\{\bar{Q}\bar{P}\} \\
&= \lambda_i\{(X_\infty^{-1} + \beta^2 Y_\infty)^{-1}\beta^2 Y_\infty\} \\
&= \lambda_i\{(I + \beta^2 X_\infty Y_\infty)^{-1}\beta^2 X_\infty Y_\infty\} \\
&= \frac{\beta^2 \lambda_i\{X_\infty Y_\infty\}}{1 + \beta^2 \lambda_i\{X_\infty Y_\infty\}} \\
&= \frac{\beta^2 \nu_i^2}{1 + \beta^2 \nu_i^2}.
\end{aligned}$$

□

Remark 8.4.13 Suppose $[\beta\bar{N} \;\; \bar{M}]$ is in balanced coordinates. Then its controllability Gramian $\bar{P}$ and its observability Gramian $\bar{Q}$ satisfy

$$\bar{P} = \bar{Q} = \bar{\Sigma}$$

where $\bar{\Sigma} = \text{diag}(\bar{\sigma}_1, \bar{\sigma}_2, \ldots, \bar{\sigma}_n)$ is the balanced Gramian. But then from Lemma 8.4.11,

$$Y_\infty = \beta^{-2}\bar{\Sigma}$$

and

$$X_\infty = (\bar{\Sigma}^{-1} - \bar{\Sigma})^{-1},$$

which are both diagonal, with product

$$\begin{aligned} X_\infty Y_\infty &= \beta^{-2}(\bar{\Sigma}^{-1} - \bar{\Sigma})^{-1}\bar{\Sigma} \\ &= \text{diag}(\beta^{-2}\bar{\sigma}_1^2(1-\bar{\sigma}_1^2)^{-1}, \beta^{-2}\bar{\sigma}_2^2(1-\bar{\sigma}_2^2)^{-1}, \ldots, \beta^{-2}\bar{\sigma}_n^2(1-\bar{\sigma}_n^2)^{-1}) \\ &= \text{diag}(\nu_1^2, \nu_2^2, \ldots, \nu_n^2) \\ &= N^2, \end{aligned}$$

using Lemma 8.4.12. Thus, G may be put into $\mathcal{H}_\infty$-balanced coordinates by applying the *diagonal* state similarity transform

$$\beta^{1/2}(I - \bar{\Sigma}^2)^{-(1/4)},$$

(which was derived in a straightforward way by applying the method of [26, Section 4]).

Remark 8.4.14 Suppose G is in $\mathcal{H}_\infty$-balanced coordinates. Then

$$X_\infty = Y_\infty = N$$

where $N = \text{diag}(\nu_1, \nu_2, \ldots, \nu_n)$. But then from Lemma 8.4.11,

$$\bar{P} = \beta^2 N$$

and

$$\bar{Q} = (N^{-1} + \beta^2 N)^{-1},$$

which are both diagonal, with product

$$\begin{aligned} \bar{P}\bar{Q} &= \beta^2 N(N^{-1} + \beta^2 N)^{-1} \\ &= \text{diag}(\beta^2\nu_1^2(1+\beta^2\nu_1^2)^{-1}, \beta^2\nu_2^2(1+\beta^2\nu_2^2)^{-1}, \ldots, \beta^2\nu_n^2(1+\beta^2\nu_n^2)^{-1}) \\ &= \text{diag}(\bar{\sigma}_1^2, \bar{\sigma}_2^2, \ldots, \bar{\sigma}_n^2) \\ &= \bar{\Sigma}^2, \end{aligned}$$

using Lemma 8.4.12. Thus, $[\beta\bar{N} \;\; \bar{M}]$ may be put into balanced coordinates by applying the *diagonal* state similarity transform

$$\beta^{-(1/2)}(I + \beta^2 N^2)^{-(1/4)},$$

(which was derived in a straightforward way by applying the method of [26, Section 4]).

8.4.4 Model Reduction via the Coprime Factors

We are now in a position to carry out balanced truncation of the normalized left-coprime factors $[\beta\bar{N} \;\; \bar{M}]$ of the scaled plant βG, to obtained a reduced-order plant. The analysis here and in the next subsection is reminiscent of [40, Chapter 5].

Procedure 8.4.15 (Reduced-order plant by coprime factorization of βG) *Let $G = (A, B, C)$ be minimal with n-states and let $\gamma > \max\{1, \gamma_o\}$. Let $\beta = (1-\gamma^{-2})^{1/2}$. Let $\beta G = \bar{M}^{-1}\beta\bar{N}$ be the normalized left-coprime factorization of βG based on the HFARE—that is, let $[\beta\bar{N} \;\; \bar{M}] = (\bar{A}, \bar{B}, \bar{C}, \bar{D})$ be as given in Lemma 8.4.9. Let the state-space realization of $[\beta\bar{N} \;\; \bar{M}]$ be in balanced coordinates with Hankel singular values $1 > \bar{\sigma}_1 \geq \bar{\sigma}_2 \geq \cdots \geq \bar{\sigma}_n > 0$—that is, let the controllability Gramian $\bar{P}$ and the observability Gramian $\bar{Q}$ of $[\beta\bar{N} \;\; \bar{M}]$ satisfy*

$$\bar{P} = \bar{Q} = \bar{\Sigma}$$

where $\bar{\Sigma} = \operatorname{diag}(\bar{\sigma}_1, \bar{\sigma}_2, \ldots, \bar{\sigma}_n)$. Pick $k < n$ such that $\bar{\sigma}_k > \bar{\sigma}_{k+1}$ and partition $\bar{\Sigma}$ accordingly into

$$\bar{\Sigma} = \begin{bmatrix} \bar{\Sigma}_1 & 0 \\ 0 & \bar{\Sigma}_2 \end{bmatrix}$$

where $\bar{\Sigma}_1 = \operatorname{diag}(\bar{\sigma}_1, \ldots, \bar{\sigma}_k)$ and $\bar{\Sigma}_2 = \operatorname{diag}(\bar{\sigma}_{k+1}, \ldots, \bar{\sigma}_n)$. Partition $\bar{A}$, $\bar{B}$ and $\bar{C}$ conformably with the partitioning of $\bar{\Sigma}$:

$$\bar{A} = \begin{bmatrix} \bar{A}_{11} & \bar{A}_{12} \\ \bar{A}_{21} & \bar{A}_{22} \end{bmatrix}, \qquad \bar{B} = \begin{bmatrix} \bar{B}_1 \\ \bar{B}_2 \end{bmatrix} \quad \text{and} \quad \bar{C} = \begin{bmatrix} \bar{C}_1 & \bar{C}_2 \end{bmatrix}.$$

Define $[\beta\bar{N}_r \;\; \bar{M}_r] := (\bar{A}_{11}, \bar{B}_1, \bar{C}_1, \bar{D})$. Then a k-state reduced-order plant is $\bar{G}_r := \bar{M}_r^{-1}\bar{N}_r$ and [43] $\beta\bar{G}_r = \bar{M}_r^{-1}\beta\bar{N}_r$ is a normalized left-coprime factorization.

How does this reduced-order plant compare with the one obtained via $\mathcal{H}_\infty$-balanced truncation in Procedure 8.4.1? The next result answers this question.

Theorem 8.4.16 *The plant model reduction schemes described in Procedure 8.4.1 and Procedure 8.4.15 yield identical reduced-order plants. To be precise, let G_r be the k-state reduced-order plant obtained by performing $\mathcal{H}_\infty$-balanced truncation (Procedure 8.4.1) on the full-order plant G for $\gamma > \max\{1, \gamma_o\}$. Let $\bar{G}_r$ be the k-state reduced-order plant obtained by performing balanced truncation of the coprime factors $[\beta\bar{N} \;\; \bar{M}]$ of βG (Procedure 8.4.15) where $\beta = (1-\gamma^{-2})^{1/2}$. Then*

$$G_r = \bar{G}_r.$$

Proof Suppose $G = (A, B, C)$ is $\mathcal{H}_\infty$-balanced, and we form G_r by $\mathcal{H}_\infty$-balanced truncation. Thus, according to Procedure 8.4.1, we have the partitioning

$$X_\infty = Y_\infty = N = \begin{bmatrix} N_1 & 0 \\ 0 & N_2 \end{bmatrix}$$

where $N_1 = \operatorname{diag}(\nu_1, \dots, \nu_k)$ and $N_2 = \operatorname{diag}(\nu_{k+1}, \dots, \nu_n)$, together with

$$A = \begin{bmatrix} A_{11} & A_{12} \\ A_{21} & A_{22} \end{bmatrix}, \qquad B = \begin{bmatrix} B_1 \\ B_2 \end{bmatrix} \quad \text{and} \quad C = \begin{bmatrix} C_1 & C_2 \end{bmatrix}.$$

The k-state reduced-order plant G_r is then

$$G_r = \left[\begin{array}{c|c} A_{11} & B_1 \\ \hline C_1 & 0 \end{array}\right]. \tag{8.7}$$

Using Lemma 8.4.9, we can use N to construct the normalized left-coprime factorization $\beta G = \bar{M}^{-1}\beta\bar{N}$ (with the partitioning corresponding to that of N made explicit):

$$[\beta\bar{N} \;\; \bar{M}] = \left[\begin{array}{cc|cc} A_{11} - \beta^2 N_1 C_1^T C_1 & A_{12} - \beta^2 N_1 C_1^T C_2 & \beta B_1 & -\beta^2 N_1 C_1^T \\ A_{21} - \beta^2 N_2 C_2^T C_1 & A_{22} - \beta^2 N_2 C_2^T C_2 & \beta B_2 & -\beta^2 N_2 C_2^T \\ \hline C_1 & C_2 & 0 & I \end{array}\right], \tag{8.8}$$

with Gramians $\bar{P} = \beta^2 N$ and $\bar{Q} = (N^{-1} + \beta^2 N)^{-1}$. But from Remark 8.4.14, this may be put into balanced coordinates by a *diagonal* state-transformation. Denote this diagonal balancing transformation by $L = \operatorname{diag}(L_1, L_2)$, partitioned conformably with $N = \operatorname{diag}(N_1, N_2)$. (An explicit expression for the balancing transformation L is given in Remark 8.4.14 but we shall not need it.) Applying this diagonal balancing transformation to (8.8) gives

$$[\beta\bar{N} \;\; \bar{M}] =$$

$$\left[\begin{array}{cc|cc} L_1(A_{11} - \beta^2 N_1 C_1^T C_1)L_1^{-1} & L_1(A_{12} - \beta^2 N_1 C_1^T C_2)L_2^{-1} & \beta L_1 B_1 & -\beta^2 L_1 N_1 C_1^T \\ L_2(A_{21} - \beta^2 N_2 C_2^T C_1)L_1^{-1} & L_2(A_{22} - \beta^2 N_2 C_2^T C_2)L_2^{-1} & \beta L_2 B_2 & -\beta^2 L_2 N_2 C_2^T \\ \hline C_1 L_1^{-1} & C_2 L_2^{-1} & 0 & I \end{array}\right].$$

which is a balanced realization with (Remark 8.4.14) balanced Gramian $\bar{\Sigma}$. Applying Procedure 8.4.15, the k-state reduced-order plant $\bar{G}_r$ has a normalized left-coprime factorization $\beta\bar{G}_r = \bar{M}_r^{-1}\beta\bar{N}_r$ where $[\beta\bar{N}_r \;\; \bar{M}_r]$ is obtained by balanced truncation of $[\beta\bar{N} \;\; \bar{M}]$. Thus, truncating the above balanced realization to k-states,

$$\begin{aligned} [\beta\bar{N}_r \;\; \bar{M}_r] &= \left[\begin{array}{c|cc} L_1(A_{11} - \beta^2 N_1 C_1^T C_1)L_1^{-1} & \beta L_1 B_1 & -\beta^2 L_1 N_1 C_1^T \\ \hline C_1 L_1^{-1} & 0 & I \end{array}\right] \\ &= \left[\begin{array}{c|cc} A_{11} - \beta^2 N_1 C_1^T C_1 & \beta B_1 & -\beta^2 N_1 C_1^T \\ \hline C_1 & 0 & I \end{array}\right] \\ \implies \bar{G}_r = \bar{M}_r^{-1}\bar{N}_r &= \left[\begin{array}{c|c} A_{11} & B_1 \\ \hline C_1 & 0 \end{array}\right]. \end{aligned}$$

But this is precisely G_r in (8.7). □

The following corollary is immediate on taking note of Remark 8.4.3.

Corollary 8.4.17 *Let K_r be the k-state reduced-order controller obtained by performing $\mathcal{H}_\infty$-balanced truncation (Procedure 8.4.2) on $K_{ME\infty}$, where $K_{ME\infty}$ is the full-order Normalized $\mathcal{H}_\infty$ Controller for the full-order plant G. Let $\bar{K}_r$ be the k-state Normalized $\mathcal{H}_\infty$ Controller for the k-state reduced-order plant G_r (Procedure 8.4.1 or equivalently Procedure 8.4.15). Then*

$$K_r = \bar{K}_r.$$

8.4.5 Stability and Performance with the Reduced-Order Controller

We can now piece together results from the previous sections to consider the stability and performance of the closed-loop consisting of the reduced-order controller K_r with the full-order plant G (as illustrated in Figure 8.1). It is an advantage of using $\mathcal{H}_\infty$-balanced truncation that the results may be expressed in terms of *a priori* quantities only (i.e., γ and the neglected ν_i). This comes about because the model reduction error $\hat{\epsilon}$ defined by

$$\beta\hat{\epsilon} := \|[\beta\Delta_{\bar{N}} \ \ \Delta_{\bar{M}}]\|_\infty, \tag{8.9}$$

where

$$\Delta_{\bar{N}} := \bar{N} - \bar{N}_r$$
$$\Delta_{\bar{M}} := \bar{M} - \bar{M}_r,$$

may be bounded above using γ and the neglected ν_i only. To be precise, the balanced truncation error bound of Lemma 8.4.4(iii) applied to (8.9) gives an upper bound on $\beta\hat{\epsilon}$ in terms of the neglected Hankel singular values of $[\beta\bar{N} \ \ \bar{M}]$. Hence, from Proposition 8.4.12, $\beta\hat{\epsilon}$ may be bounded in terms of the $\mathcal{H}_\infty$-characteristic values of G:

$$\beta\hat{\epsilon} \le 2\,\mathrm{trace}[\bar{\Sigma}_2] = 2\sum_{i=k+1}^{n} \frac{\beta\nu_i}{\sqrt{1+\beta^2\nu_i^2}}. \tag{8.10}$$

Therefore,

$$\hat{\epsilon} \le 2\sum_{i=k+1}^{n} \frac{\nu_i}{\sqrt{1+\beta^2\nu_i^2}}, \tag{8.11}$$

and by inspection, a weaker but simpler bound is

$$\hat{\epsilon} \le 2\sum_{i=k+1}^{n} \nu_i = 2\,\mathrm{trace}[N_2]. \tag{8.12}$$

This last upper bound is 'twice the sum of the tail,' just as in ordinary balanced truncation (Lemma 8.4.4(iii)). Using either of the above upper bounds in place of $\hat{\epsilon}$ in the following results gives a (conservative) statement of stability and performance using the reduced-order controller *without* having to calculate the reduced-order controller in advance.

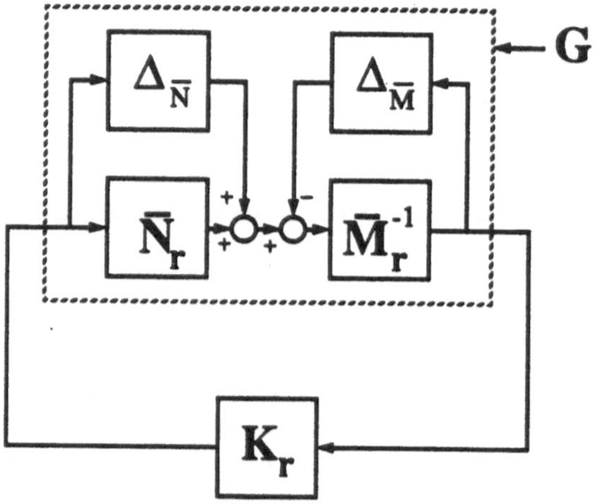

Figure 8.1: Reduced-order controller with full-order plant

Proposition 8.4.18 *Let K_r be the k-state reduced-order controller obtained by performing $\mathcal{H}_\infty$-balanced truncation (Procedure 8.4.2) on the full-order Normalized $\mathcal{H}_\infty$ Controller for the full-order plant G with $\gamma > \max\{1, \gamma_o\}$. Let $\beta G = \bar{M}^{-1}\beta\bar{N}$ be the normalized left-coprime factorization of βG given in Lemma 8.4.9. Let $\beta G_r = \bar{M}_r^{-1}\beta\bar{N}_r$ be the normalized left-coprime factorization (Procedure 8.4.15) of the k-state reduced-order plant G_r obtained by performing $\mathcal{H}_\infty$-balanced truncation (Procedure 8.4.1). Let*

$$K_r = U_r V_r^{-1}$$

be any right-coprime factorization of K_r. Define

$$R_{rr} := \bar{M}_r V_r - \bar{N}_r U_r.$$

Let $\hat{\epsilon}$ be the model reduction error as defined in (8.9) above. Then the condition

$$\hat{\epsilon}\left\| \begin{bmatrix} \beta V_r \\ U_r \end{bmatrix} R_{rr}^{-1} \right\|_\infty < 1$$

ensures closed-loop stability of the system consisting of the reduced-order controller K_r and the full-order plant G.

Proof The proof is basically an application of the Small Gain Theorem. Rather than repeat the argument here, we will for brevity quote a result [40, Corollary 3.7] which, for our case, states that K_r stabilizes G if K_r stabilizes G_r (which it does by construction) and

$$\hat{\epsilon}\left\|\begin{bmatrix} \beta S_{rr}\bar{M}_r^{-1} \\ K_r S_{rr}\bar{M}_r^{-1} \end{bmatrix}\right\|_\infty < 1$$

where $S_{rr} := (I - G_r K_r)^{-1}$. The claim of the proposition follows easily on making the necessary substitutions to obtain

$$S_{rr} = V_r R_{rr}^{-1} \bar{M}_r \tag{8.13}$$

and then

$$\begin{bmatrix} \beta S_{rr}\bar{M}_r^{-1} \\ K_r S_{rr}\bar{M}_r^{-1} \end{bmatrix} = \begin{bmatrix} \beta V_r \\ U_r \end{bmatrix} R_{rr}^{-1}.$$

□

Corollary 8.4.17 shows that K_r is in fact the normalized $\mathcal{H}_\infty$ controller for G_r. The associated closed-loop transfer function is

$$H_{rr} := \begin{bmatrix} S_{rr}G_r & S_{rr}G_r K_r \\ K_r S_{rr} G_r & K_r S_r r \end{bmatrix}$$

where $S_{rr} := (I - G_r K_r)^{-1}$. Define

$$\hat{\gamma} := \|H_{rr}\|_\infty \tag{8.14}$$

to be the actual $\mathcal{H}_\infty$-norm of H_{rr}. Then

$$\hat{\gamma} < \gamma \tag{8.15}$$

is immediate, and the following corollary to Proposition to 8.4.18 is obtained.

Corollary 8.4.19 *With definitions as above, we have that*

$$\left\|\begin{bmatrix} \beta V_r \\ U_r \end{bmatrix} R_{rr}^{-1}\right\|_\infty \le \beta + \hat{\gamma}, \tag{8.16}$$

so the condition

$$\hat{\epsilon}(\beta + \hat{\gamma}) < 1$$

is sufficient to ensure closed-loop stability of the system consisting of the reduced-order controller K_r and the full-order plant G.

Proof Appendix A.6. □

Remark 8.4.20 An *a priori* test for the stability of the system consisting of K_r with G is immediate. Just replace $\hat{\gamma}$ with its upper bound γ (equations (8.14) and (8.15)), and replace $\hat{\epsilon}$ with ϵ, where ϵ is an upper bound on $\hat{\epsilon}$, such as that given in (8.11). It is simple to verify that

$$\epsilon(\beta+\gamma) < 1 \quad \Longrightarrow \quad \hat{\epsilon}(\beta+\hat{\gamma}) < 1.$$

Hence to test if K_r stabilizes G it suffices to check if $\epsilon(\beta+\gamma) < 1$, a simple test which depends only on γ and the truncated ν_i.

Now assume that $\hat{\epsilon}(\beta+\hat{\gamma}) < 1$ so that, by Corollary 8.4.19, the closed-loop consisting of the reduced-order controller K_r and the full-order plant G is stable. From Remark 7.3.1 we know that the bound $\|H_{ME\infty}\|_\infty < \gamma$ inherent in Normalized $\mathcal{H}_\infty$ Problem gives robust stability guarantees, where

$$H_{ME\infty} = \begin{bmatrix} SG & SGK_{ME\infty} \\ K_{ME\infty}SG & K_{ME\infty}S \end{bmatrix},$$

is the closed-loop transfer function associated with G and its Normalized $\mathcal{H}_\infty$ Controller $K_{ME\infty}$, and where $S = (I - GK_{ME\infty})^{-1}$. The next proposition tells us how much the bound $\|H_{ME\infty}\|_\infty < \gamma$ is degraded (increased) by using the reduced-order controller K_r in place of the full-order Normalized $\mathcal{H}_\infty$ Controller $K_{ME\infty}$. For convenience, define the associated closed-loop

$$H_r := \begin{bmatrix} S_rG & S_rGK_r \\ K_rS_rG & K_rS_r \end{bmatrix}$$

where $S_r := (I - GK_r)^{-1}$; that is, the closed-loop transfer function when K_r is used in place of $K_{ME\infty}$.

Proposition 8.4.21 *Definitions as above. Assume* $\hat{\epsilon}(\beta+\hat{\gamma}) < 1$ *so that the reduced-order controller* K_r *stabilizes the full-order plant* G. *Then the associated closed-loop transfer function satisfies*

$$\|H_r\|_\infty \le \hat{\gamma} + \frac{\hat{\epsilon}(1+\hat{\gamma})(1+\beta+\hat{\gamma})}{(1-\hat{\epsilon}(\beta+\hat{\gamma}))}. \tag{8.17}$$

Proof Appendix A.6. □

Remark 8.4.22 An *a priori* upper bound on $\|H_r\|_\infty$ is immediate from (8.17). Just replace $\hat{\gamma}$ with its upper bound γ (equations (8.14) and (8.15)), and replace $\hat{\epsilon}$ with ϵ, where ϵ is an upper bound on $\hat{\epsilon}$ such as that given in (8.11). Provided that $\epsilon(\beta+\gamma) < 1$, Remark 8.4.20 predicts that K_r stabilizes G. Under this condition, it is simple to verify that the right-hand side of (8.17) can only increase under the transformation $\hat{\epsilon} \to \epsilon$, $\hat{\gamma} \to \gamma$. That is,

$$\|H_r\|_\infty < \gamma + \frac{\epsilon(1+\gamma)(1+\beta+\gamma)}{(1-\epsilon(\beta+\gamma))}, \tag{8.18}$$

which depends only on γ and the truncated ν_i.

Remark 8.4.23 Note that as γ increases, the *a posteriori* result of Proposition 8.4.21 remains viable, whereas the *a priori* result of Remark 8.4.22 becomes increasingly weak. In the limit as $\gamma \to \infty$ (the LQG case), Proposition 8.4.21 is still viable, whereas Remark 8.4.22 becomes vacuous.

Remark 8.4.24 The upper bound (8.17) is tight, in that as the model reduction error $\hat{\epsilon} \to 0$, so the proposition recovers the definition $\hat{\gamma} = \|H_{rr}\|_\infty$. Similarly, the upper bound (8.18) is tight, in that as the model reduction error bound $\epsilon \to 0$, so the full-order bound $\|H_{ME\infty}\|_\infty < \gamma$ is recovered.

8.4.6 A Numerical Example

The results derived in the previous subsection are illustrated here using a numerical example. Consider the system

$$G = \left[\begin{array}{c|c} A & B \\ \hline C & 0 \end{array}\right] = \left[\begin{array}{cc|c} 3/4 & -98/201 & 1 \\ -98/201 & -9999/200 & 1 \\ \hline 1 & 1 & 0 \end{array}\right].$$

This system was constructed in [35, Section VII] to be in LQG-balanced coordinates with LQG-characteristic values $\mu_1 = 2.0000$ and $\mu_2 = 0.0100$. To verify this, one merely has to check that $M = \mathrm{diag}(\mu_1, \mu_2)$ is the stabilizing solution of the CARE and FARE. The system has a left-half plane zero at -24.1349 and poles at 0.7547 and -49.9997. Using γ-iteration gives $\gamma_o = 2.3559$. We chose seven representative values of $\gamma > \gamma_o$ at which to perform $\mathcal{H}_\infty$-balanced truncation: $\gamma = 2.36$, 3, 7, 10, 40, 100 and ∞.

The results are tabulated in Figures 8.2 and 8.3: the 'predicted' results in Figure 8.2 and the 'actual' results in Figure 8.3. We know that when $\gamma \to \infty$, the $\mathcal{H}_\infty$-balancing method recovers the LQG-balancing method. This is confirmed in Figures 8.2 and 8.3, where the last row (corresponding to $\gamma \to \infty$) of Columns (a), (b) and (i) agrees with the results of [35, Section VII].

For each of the chosen values of γ, the HCARE and HFARE are solved and the $\mathcal{H}_\infty$-characteristic values are calculated using Proposition 8.3.1. This gives Columns (a) and (b) of Figure 8.2. For each choice of γ we have $\nu_1 \gg \nu_2$, which prompts us to consider obtaining single-state reduced-order controllers by discarding ν_2 via $\mathcal{H}_\infty$-balanced truncation (Procedure 8.4.2). But before calculating the reduced-order controllers, we can predict their success using the *a priori* data γ and ν_2. Using (8.11), an upper bound on the model reduction error $\hat{\epsilon}$ may be calculated from the neglected ν_2, giving Column (c). As explained in Remark 8.4.20, this upper bound on $\hat{\epsilon}$ may then be used in place of $\hat{\epsilon}$ in Corollary 8.4.19 to predict stability of each reduced-order controller K_r with the full-order plant G, Column (d). Notice that stability of K_r with G is predicted for all but the two largest choices of γ, when nothing can be deduced. In the cases where stability of K_r with G is predicted, the performance of that configuration can be bounded using Remark 8.4.22, as shown in Column (e).

γ	$\beta = \sqrt{1-\gamma^{-2}}$	(a) ν_1	(b) ν_2	(c) Upper bound on $\hat{\epsilon}$ from (8.11)	(d) Does Remark 8.4.20 $\Rightarrow K_r$ stabilises G?	(e) Upper bound on $\|H_r\|_\infty$ from Remark 8.4.22
2.3600	0.9058	2.3544	0.0100	0.0200	Yes	2.6667
3.0000	0.9428	2.2030	0.0100	0.0200	Yes	3.4293
7.0000	0.9897	2.0339	0.0100	0.0200	Yes	8.7119
10.0000	0.9950	2.0167	0.0100	0.0200	Yes	13.3828
40.0000	0.9997	2.0010	0.0100	0.0200	Yes	231.3253
100.0000	0.9999	2.0002	0.0100	0.0200	No	———
∞	1.0000	2.0000	0.0100	0.0200	No	———

Figure 8.2: $\mathcal{H}_\infty$-balanced truncation example: *a priori* numerical results

γ	(f) $\hat{\gamma}$	(g) Actual value of $\hat{\epsilon}$ from (8.9)	(h) Does Cor. 8.4.19 $\Rightarrow K_r$ stabilises G?	(i) Does K_r stabilise G?	(j) Upper bound on $\|H_r\|_\infty$ from Prop. 8.4.21	(k) Actual value of $\|H_r\|_\infty$	(l) Actual value of $\|H_{MB\infty}\|_\infty$
2.3600	2.3600	0.0182	Yes	Yes	2.6373	2.4273	2.3600
3.0000	2.9076	0.0189	Yes	Yes	3.2939	3.0249	2.9087
7.0000	4.1396	0.0198	Yes	Yes	4.8339	4.4268	4.1524
10.0000	4.3451	0.0199	Yes	Yes	5.0997	4.6688	4.3612
40.0000	4.5472	0.0200	Yes	Yes	5.3642	4.9092	4.5669
100.0000	4.5590	0.0200	Yes	Yes	5.3794	4.9233	4.5790
∞	4.5612	0.0200	Yes	Yes	5.3823	4.9260	4.5812

Figure 8.3: $\mathcal{H}_\infty$-balanced truncation example: exact and *a posteriori* numerical results

Having predicted the satisfactory behaviour of the reduced-order controllers we can go ahead and calculate them using Procedure 8.4.2, from which Figure 8.3 is constructed. The model reduction error $\hat{\epsilon}$ may be calculated exactly from (8.9) using the algorithm in [10], to give Column (g). Stability of K_r with G may be tested for explicitly by calculating the closed-loop poles, to give Column (i). The actual $\mathcal{H}_\infty$-norm of the closed-loop of K_r with G is also calculated explicitly using the algorithm in [10], to give Column (k). Likewise, $\hat{\gamma}$, the actual $\mathcal{H}_\infty$-norm of the closed-loop of K_r with G_r is calculated explicitly to give Column (f). For reference, in Column (l) we give the full-order results, that is, the actual $\mathcal{H}_\infty$-norm of the closed-loop $H_{ME\infty}$ consisting of the full-order controller $K_{ME\infty}$ with G.

Using the *a posteriori* data $\hat{\gamma}$ and $\hat{\epsilon}$, Corollary 8.4.19 may be used to check if stability of K_r with G is predicted, see Column (h). Proposition 8.4.21 then allows calculation of an upper bound on the $\mathcal{H}_\infty$-norm of the closed-loop of K_r with G (Column (j)). Notice that it is predicted that K_r stabilizes G for all the values of γ, including infinity. Also, the upper bound in Column (j) is tight (within 10% of the actual value in Column (k)), again even for large γ.

Comparing the two tables of results it is clear that the $\mathcal{H}_\infty$-balanced truncation method can give a good reduced-order controller. Furthermore, when γ is small (of the order of γ_o), indicating a preference for robustness over LQG performance, the $\mathcal{H}_\infty$-balanced truncation method gives accurate *a priori* prediction of the performance of K_r with G. This is particularly clear on comparing the first three rows of the two tables. With the *a posteriori* data $\hat{\gamma}$ and $\hat{\epsilon}$, even tighter results are possible; they indicate that a good reduced-order controller can be obtained even when γ is large (in which case the *a priori* predictions may be weak, as indicated in Remark 8.4.23). This is particularly clear on comparing the final three rows of the two tables.

Chapter 9

LQG and $\mathcal{H}_\infty$ Monotonicity

9.1 Introduction

The aim of this chapter is to explore two properties of minimum entropy $\mathcal{H}_\infty$ controllers which we shall call *LQG monotonicity* and *$\mathcal{H}_\infty$ monotonicity.* Recall the Minimum Entropy $\mathcal{H}_\infty$ Control Problem of Chapter 3.2.6; in particular, recall Section 3.6.1. There it was shown that the minimum entropy closed-loop $H_{ME\infty}$ satisfies (Proposition 3.6.1)

(i) $\|H_{ME\infty}\|_\infty < \gamma$.

(ii) $C(H_{ME\infty}) \leq I(H_{ME\infty};\gamma;\infty)$.

Here $C(H_{ME\infty})$ is the LQG cost (Definition 2.4.1) of the minimum entropy closed-loop and $I(H_{ME\infty};\gamma;\infty)$ is its entropy at infinity. Theorem 3.6.2 states that the upper bounds on the right-hand sides of (i) and (ii) above trade off: $I(H_{ME\infty};\gamma;\infty)$ is a monotonically decreasing function of γ. Computational evidence points towards a much stronger tradeoff: that the left-hand side of (i) is a monotonically increasing function of γ, and that the left-hand side of (ii) is a monotonically decreasing function of γ. We state this formally as two conjectures.

Conjecture 9.1.1 (LQG monotonicity) *Suppose $K_{ME\infty}$ solves the Minimum Entropy $\mathcal{H}_\infty$ Control Problem at infinity (Problem 3.2.6) for a given standard plant P and $\gamma > \gamma_o$. Then the minimum entropy closed-loop $H_{ME\infty} = \mathcal{F}(P, K_{ME\infty})$ exhibits* LQG monotonicity, *that is, the achieved LQG cost $C(H_{ME\infty})$ is a* monotonically decreasing *function of γ.*

Conjecture 9.1.2 ($\mathcal{H}_\infty$ monotonicity) *Suppose $K_{ME\infty}$ solves the Minimum Entropy $\mathcal{H}_\infty$ Control Problem at infinity (Problem 3.2.6) for a given standard plant P and $\gamma > \gamma_o$. Then the minimum entropy closed-loop $H_{ME\infty} = \mathcal{F}(P, K_{ME\infty})$ exhibits* $\mathcal{H}_\infty$ monotonicity, *that is, the achieved $\mathcal{H}_\infty$-norm $\|H_{ME\infty}\|_\infty$ is a* monotonically increasing *function of γ.*

See Figures 7.3 and 7.4 for an example of LQG and $\mathcal{H}_\infty$ monotonicity. The proof, or otherwise, of the above conjectures in the general case is currently an open problem. In the remainder of this chapter we will focus on LQG monotonicity. Our approach will be to derive a formula for the derivative of the LQG cost with respect to γ. Proof of LQG monotonicity then reduces to proving that this derivative is non-positive.

9.2 The LQG Cost and its Derivative

The first step in deriving the derivative of the LQG cost with respect to γ is to use the method of Section 4.2 to reduce the Minimum Entropy $\mathcal{H}_\infty$ Control Problem to the equivalent Minimum Entropy $\mathcal{H}_\infty$ Distance Problem (Problem 4.2.1), in both cases with entropy evaluated at infinity. Using the terminology of Chapter 3, let P be a given standard plant and let $\gamma > \gamma_o$; let $K_{ME\infty}$ be the controller which solves the associated

Minimum Entropy $\mathcal{H}_\infty$ Control Problem at infinity and let $H_{ME\infty} = \mathcal{F}(P, K_{ME\infty})$ be the corresponding minimum entropy closed-loop. Then using the terminology of Chapter 4, let $E_{ME\infty}$ be the minimum entropy error system for the Minimum Entropy $\mathcal{H}_\infty$ Distance Problem at infinity associated with the above Minimum Entropy $\mathcal{H}_\infty$ Control Problem. The error system $E_{ME\infty}$ has the structure

$$
E_{ME\infty} = \begin{bmatrix} R_{11} & R_{12} \\ R_{21} & R_{22} + \hat{Q}_{ME\infty} \end{bmatrix},
$$

where

$$
R = \begin{array}{c} p_1 - m_2 \updownarrow \\ p_2 \updownarrow \end{array} \overset{\overset{m_1 - p_2}{\longleftrightarrow} \overset{m_2}{\longleftrightarrow}}{\begin{bmatrix} R_{11} & R_{12} \\ R_{21} & R_{22} \end{bmatrix}}, \qquad R^* \in \mathcal{RH}_\infty,
$$

is determined by the standard plant P only, whilst $\hat{Q}_{ME\infty} \in \mathcal{RH}_\infty$ and minimizes the entropy over the class of error systems satisfying the $\mathcal{L}_\infty$-norm bound $\|E\|_\infty < \gamma$. Then we have

$$
\|H_{ME\infty}\|_\infty = \|E_{ME\infty}\|_\infty,
$$
$$
I(H_{ME\infty}; \gamma; \infty) = I(E_{ME\infty}; \gamma; \infty),
$$
$$
C(H_{ME\infty}) = \|H_{ME\infty}\|_2^2 = \|E_{ME\infty}\|_2^2.
$$

These relations hold because (Section 4.2) $\mathcal{F}(P, K_{ME\infty}) = U E_{ME\infty} V$ for suitable transfer function matrices U and V satisfying $U^*U = I$ and $VV^* = I$: the $\mathcal{L}_\infty$-norm, entropy, LQG cost and $\mathcal{L}_2$-norm are all unitarily invariant. We assume that $R(\infty) = 0$—this is sufficient for a finite value for the minimum entropy and LQG cost, as noted in Section 4.5.

We shall need the following state-space formula for $E_{ME\infty}$ from Theorem 4.5.2. Suppose

$$
R = \begin{array}{c} 2n \updownarrow \\ p_1 - m_2 \updownarrow \\ m_2 \updownarrow \end{array} \overset{\overset{2n}{\longleftrightarrow} \overset{m_1 - p_2}{\longleftrightarrow} \overset{p_2}{\longleftrightarrow}}{\left[\begin{array}{c|cc} A & B_1 & B_2 \\ \hline C_1 & 0 & 0 \\ C_2 & 0 & 0 \end{array} \right]} =: \left[\begin{array}{c|c} A & B \\ \hline C & 0 \end{array} \right]
$$

is minimal and $R^* \in \mathcal{RH}_\infty$ i.e., $-A$ is asymptotically stable. Then the minimum entropy error system $E_{ME\infty}$ is

$$
E_{ME\infty} = \begin{bmatrix} R_{11} & R_{12} \\ R_{21} & R_{22} + \hat{Q}_{ME\infty} \end{bmatrix}
$$

where $\hat{Q}_{ME\infty} \in \mathcal{RH}_\infty$ has a realization

$$
\hat{Q}_{ME\infty} := \left[\begin{array}{c|c} \hat{A} & \hat{B} \\ \hline \hat{C} & 0 \end{array} \right] = \left[\begin{array}{c|c} -A^T - \gamma^{-2} Y Z^{-1} B_2 B_2^T - \gamma^{-2} C_1^T C_1 X & \gamma^{-1} Y Z^{-1} B_2 \\ \hline -\gamma^{-1} C_2 X & 0 \end{array} \right] \quad (9.1)
$$

and where $X = X^T \leq 0$ solves

$$XA^T + AX + \gamma^{-2}XC_1^TC_1X + BB^T = 0 \tag{9.2}$$

such that

$$-(A + \gamma^{-2}XC_1^TC_1) \qquad \text{is asymptotically stable,} \tag{9.3}$$

where $Y = Y^T \leq 0$ solves

$$YA + A^TY + \gamma^{-2}YB_1B_1^TY + C^TC = 0 \tag{9.4}$$

such that

$$-(A + \gamma^{-2}B_1B_1^TY) \qquad \text{is asymptotically stable,} \tag{9.5}$$

and where

$$Z := \gamma^{-2}XY - I.$$

Let $\hat{Q}_{ME\infty}$ have controllability Gramian $\hat{\mathcal{P}}$ and observability Gramian $\hat{\mathcal{Q}}$. Then (since $\hat{Q}_{ME\infty} \in \mathcal{RH}_\infty$) $\hat{\mathcal{P}} = \hat{\mathcal{P}}^T \geq 0$ is the unique solution of the Lyapunov equation

$$\hat{\mathcal{P}}\hat{A}^T + \hat{A}\hat{\mathcal{P}} + \hat{B}\hat{B}^T = 0, \tag{9.6}$$

and $\hat{\mathcal{Q}} = \hat{\mathcal{Q}}^T \geq 0$ is the unique solution of the Lyapunov equation

$$\hat{\mathcal{Q}}\hat{A} + \hat{A}^T\hat{\mathcal{Q}} + \hat{C}^T\hat{C} = 0. \tag{9.7}$$

Similarly, let $R(-s)$ have controllability Gramian $-\mathcal{X}$ and observability Gramian $-\mathcal{Y}$. Then (since $R(-s) \in \mathcal{RH}_\infty$ and R is minimal) $\mathcal{X} = \mathcal{X}^T < 0$ is the unique solution of the Lyapunov equation

$$\mathcal{X}A^T + A\mathcal{X} + BB^T = 0,$$

and $\mathcal{Y} = \mathcal{Y}^T < 0$ is the unique solution of the Lyapunov equation

$$\mathcal{Y}A + A^T\mathcal{Y} + C^TC = 0.$$

Note carefully that $\mathcal{X}$ and $\mathcal{Y}$ are independent of γ but $\hat{\mathcal{P}}$ and $\hat{\mathcal{Q}}$ are not.

An expression for $\|E_{ME\infty}\|_2^2$ (= $C(H_{ME\infty})$ the LQG cost) may now be stated.

Lemma 9.2.1 *Consider the Minimum Entropy $\mathcal{H}_\infty$ Distance Problem at infinity as described above. Let $E_{ME\infty}$ be the minimum entropy error system. Then*

$$\|E_{ME\infty}\|_2^2 = -\operatorname{trace}[C\mathcal{X}C^T] + \operatorname{trace}[\hat{C}\hat{\mathcal{P}}\hat{C}^T].$$

Proof Since $R^* \in \mathcal{RH}_\infty$ and $\hat{Q}_{ME\infty} \in \mathcal{RH}_\infty$ with $R(\infty) = 0$ and $\hat{Q}_{ME\infty}(\infty) = 0$, we have $R^* \in \mathcal{RH}_2$ and $\hat{Q}_{ME\infty} \in \mathcal{RH}_2$. Hence R and $\hat{Q}_{ME\infty}$ are orthogonal in the Hilbert space $\mathcal{RL}_2$. It follows that

$$\begin{aligned}
\|E_{ME\infty}\|_2^2 &= \|R\|_2^2 + \|\hat{Q}_{ME\infty}\|_2^2 \\
&= \|R^*\|_2^2 + \|\hat{Q}_{ME\infty}\|_2^2 \\
&= \|R(-s)\|_2^2 + \|\hat{Q}_{ME\infty}\|_2^2 \\
&= \operatorname{trace}[C(-\mathcal{X})C^T] + \operatorname{trace}[\hat{C}\hat{\mathcal{P}}\hat{C}^T],
\end{aligned}$$

where the last line is obtained using the standard evaluation of the $\mathcal{H}_2$-norm of a system in terms of its controllability Gramian. □

The next result allows us to calculate the derivative of $\|E_{ME\infty}\|_2^2$ with respect to γ, without needing to differentiate $\hat{\mathcal{P}}$ or $\hat{\mathcal{Q}}$.

Theorem 9.2.2 *Definitions and assumptions as above. Then*

(i)

$$\frac{d}{d\gamma}(\|E_{ME\infty}\|_2^2) = 2\,\text{trace}[\frac{d\hat{A}}{d\gamma}\hat{\mathcal{P}}\hat{\mathcal{Q}} + \frac{d\hat{B}}{d\gamma}\hat{B}^T\hat{\mathcal{Q}} + \hat{\mathcal{P}}\hat{C}^T\frac{d\hat{C}}{d\gamma}]$$

(ii) *Define* $\bar{X} := 2\gamma^{-1}X - dX/d\gamma$ *and* $\bar{Y} := 2\gamma^{-1}Y - dY/d\gamma$. *Then*

$$\begin{aligned}
\frac{d\hat{A}}{d\gamma} &= -\gamma^{-2}Z^{-T}\bar{Y}Z^{-1}B_2B_2^T + \gamma^{-2}C_1^TC_1\bar{X} + \gamma^{-4}YZ^{-1}\frac{dX}{d\gamma}YZ^{-1}B_2B_2^T,\\
\frac{d\hat{B}}{d\gamma} &= \gamma^{-1}Z^{-T}\bar{Y}Z^{-1}B_2 + \gamma^{-2}YZ^{-1}B_2 - \gamma^{-3}YZ^{-1}\frac{dX}{d\gamma}YZ^{-1}B_2,\\
\frac{d\hat{C}}{d\gamma} &= \gamma^{-2}C_2X - \gamma^{-1}C_2\frac{dX}{d\gamma}.
\end{aligned}$$

Proof *Part (i)* We follow the method used in [61, Appendix 10.1] in a different context. Using Lemma 9.2.1 we have

$$\begin{aligned}
\frac{d}{d\gamma}(\|E_{ME\infty}\|_2^2) &= \frac{d}{d\gamma}\left(-\,\text{trace}[C\mathcal{X}C^T] + \text{trace}[\hat{C}\hat{\mathcal{P}}\hat{C}^T]\right)\\
&= \text{trace}[\frac{d\hat{\mathcal{P}}}{d\gamma}\hat{C}^T\hat{C} + \hat{\mathcal{P}}\frac{d}{d\gamma}(\hat{C}^T\hat{C})]\\
&= -2\,\text{trace}[\frac{d\hat{\mathcal{P}}}{d\gamma}\hat{\mathcal{Q}}\hat{A}] + \text{trace}[\hat{\mathcal{P}}\frac{d}{d\gamma}(\hat{C}^T\hat{C})] \qquad (9.8)
\end{aligned}$$

on substituting for $\hat{C}^T\hat{C}$ from (9.7). Next, differentiate (9.6), postmultiply by $\hat{\mathcal{Q}}$ and take the trace to get

$$-2\,\text{trace}[\frac{d\hat{\mathcal{P}}}{d\gamma}\hat{\mathcal{Q}}\hat{A}] = 2\,\text{trace}[\frac{d\hat{A}}{d\gamma}\hat{\mathcal{P}}\hat{\mathcal{Q}}] + \text{trace}[\frac{d}{d\gamma}(\hat{B}\hat{B}^T)\hat{\mathcal{Q}}].$$

Substitute this into (9.8) and use the easily verified fact that

$$\begin{aligned}
\frac{d}{d\gamma}(\hat{B}\hat{B}^T) &= \frac{d\hat{B}}{d\gamma}\hat{B}^T + \left(\frac{d\hat{B}}{d\gamma}\hat{B}^T\right)^T\\
\frac{d}{d\gamma}(\hat{C}^T\hat{C}) &= \hat{C}^T\frac{d\hat{C}}{d\gamma} + \left(\hat{C}^T\frac{d\hat{C}}{d\gamma}\right)^T.
\end{aligned}$$

The result follows on collecting terms.

Part (ii) Appendix A.7. □

The following corollary provides the desired expression for the derivative of $\|E_{ME\infty}\|_2^2$ (which equals the derivative of the LQG cost of the minimum entropy $\mathcal{H}_\infty$ control system).

Corollary 9.2.3

$$\begin{aligned}\frac{d}{d\gamma}(\|E_{ME\infty}\|_2^2) &= 2\gamma^{-2}\,\mathrm{trace}[Z^{-T}\bar{Y}Z^{-1}B_2B_2^T(YZ^{-1}-\hat{\mathcal{P}})\hat{\mathcal{Q}}] \\ &\quad - 2\gamma^{-4}\,\mathrm{trace}[Z^{-T}Y\frac{dX}{d\gamma}YZ^{-1}B_2B_2^T(YZ^{-1}-\hat{\mathcal{P}})\hat{\mathcal{Q}}] \\ &\quad + 2\gamma^{-2}\,\mathrm{trace}[C_1^TC_1\bar{X}\hat{\mathcal{P}}\hat{\mathcal{Q}}] + 2\gamma^{-2}\,\mathrm{trace}[\hat{\mathcal{P}}XC_2^TC_2\frac{dX}{d\gamma}].\end{aligned}$$

Proof Immediate on substitution of Part (ii) of Theorem 9.2.2 into Part (i): collect terms and use $\mathrm{trace}[\hat{\mathcal{P}}\hat{C}^T\hat{C}] = \mathrm{trace}[\hat{B}\hat{B}^T\hat{\mathcal{Q}}]$ which follows from (9.6) and (9.7). □

9.3 On LQG Monotonicity

LQG monotonicity is equivalent to $d(\|E_{ME\infty}\|_2^2)/d\gamma \le 0$. In order to deduce anything about the sign of the expression for $d(\|E_{ME\infty}\|_2^2)/d\gamma$ in Corollary 9.2.3 above, we need to know more about the sign definiteness of its constituent matrices. We certainly know that $X \le 0$ and $Y \le 0$; that $\hat{\mathcal{P}} \ge 0$ and $\hat{\mathcal{Q}} \ge 0$; that $\lambda_i\{Z\} < 0$. The next lemma deals with the remaining matrices.

Lemma 9.3.1 *Let X, Y, Z, $\hat{\mathcal{P}}$ and $\hat{\mathcal{Q}}$ be as defined in the previous section. Then*

(i) $dX/d\gamma \ge 0$ *and* $dY/d\gamma \ge 0$.

(ii) $\bar{X} \le 0$ *and* $\bar{Y} \le 0$.

(iii) $YZ^{-1} - \hat{\mathcal{P}} \ge 0$.

Proof *Part (i)* Differentiate equation (9.2) with respect to γ (valid by [50]) to get

$$\frac{dX}{d\gamma}A_X^T + A_X\frac{dX}{d\gamma} + 2\gamma^{-3}XC_1^TC_1X = 0, \tag{9.9}$$

where

$$A_X := -(A + \gamma^{-2}XC_1^TC_1)$$

is asymptotically stable from (9.3). Applying a standard result on Lyapunov equations [26, Theorem 3.3(7)] to (9.9) gives $dX/d\gamma \ge 0$, as claimed. A similar method based on differentiating (9.4) and using (9.5) leads to $dY/d\gamma \ge 0$.

Part (ii) Immediate from Part (i), the definition of $\bar{X}$ and $\bar{Y}$ in Theorem 9.2.2 and the fact that $X \leq 0$ and $Y \leq 0$.

Part (iii) It is first claimed that

$$YZ^{-1}\hat{A}^T + \hat{A}YZ^{-1} + \hat{B}\hat{B}^T + Z^{-T}C_2^TC_2Z^{-1} + C_1^TC_1 = 0. \tag{9.10}$$

This is easily shown using (9.2), (9.4) and the expressions for $\hat{A}$ and $\hat{B}$ from (9.1). Subtract (9.6) from (9.10):

$$(YZ^{-1} - \hat{\mathcal{P}})\hat{A}^T + \hat{A}(YZ^{-1} - \hat{\mathcal{P}}) + Z^{-T}C_2^TC_2Z^{-1} + C_1^TC_1 = 0.$$

This is a Lyapunov equation, and $\hat{A}$ is asymptotically stable. Hence, using [26, Theorem 3.3(7)] again, it follows that $(YZ^{-1} - \hat{\mathcal{P}}) \geq 0$. □

If $M = M^T \geq 0$ and $N = N^T \geq 0$ are arbitrary nonnegative definite symmetric matrices then

$$\text{trace}[MN] = \text{trace}[M^{1/2}NM^{1/2}] \geq 0$$

follows. However, this result does *not* extend to three or more arbitrary nonnegative definite symmetric matrices, as the following example shows. Let

$$L = \begin{bmatrix} 101 & -10 \\ -10 & 1 \end{bmatrix}, \quad M = \begin{bmatrix} 2 & 1 \\ 1 & 1 \end{bmatrix} \quad \text{and} \quad N = \begin{bmatrix} 1 & 0 \\ 0 & 100 \end{bmatrix}.$$

Then $L > 0$, $M > 0$ and $N > 0$ but

$$\text{trace}[LMN] = -708 \not\geq 0.$$

Thus it is not clear how to deduce nonpositivity of $d(\|E_{ME\infty}\|_2^2)/d\gamma$ in Corollary 9.2.3 from knowledge of the definiteness of the constituent matrices alone. (See [6] for more on positivity of products of positive definite matrices.) We can only say with certainty that LQG monotonicity holds whenever R has only one state—for then all the terms in Corollary 9.2.3 are scalar and the nonpositivity is obvious by inspection.

Appendix A
Proof of Results Needed in the Text

A.1 Outline

Proofs of various results are stated here, rather than in the main body of the text, to preserve continuity of exposition.

A.2 A Lemma

The following technical lemma will be needed in various places—for example, in the proof of Propositions 2.3.1 and 2.3.2 and in the proof of Theorems 2.4.4 and 3.4.2.

Lemma A.2.1 *Let M be a real square matrix, let N be a real or complex matrix, and let $\epsilon \in \mathbb{R}$. Then*

(i) $-\ln\det(I-\epsilon M) = \epsilon\,\mathrm{trace}[M] + O(\epsilon^2)$.

(ii) $-\ln\det(I-\epsilon^2 N^* N) \geq \epsilon^2\,\mathrm{trace}[N^* N]$.

Proof *Part (i)* Use the Faddeev formula (see [25, Vol. 1, p88]) to obtain

$$\det(I-\epsilon M) = 1-\epsilon\,\mathrm{trace}[M] + O(\epsilon^2)$$

and expand the logarithm of this as a power series.

Part (ii) Using the well-known inequality $-\ln(1-x^2) \geq x^2$ (for $|x|<1$), we have

$$\begin{aligned} -\ln\det(I-\epsilon^2 N^* N) &= -\sum_i \ln(1-\epsilon^2\lambda_i\{N^* N\}) \\ &\geq \sum_i \epsilon^2\lambda_i\{N^* N\} \\ &= \epsilon^2\,\mathrm{trace}[N^* N], \end{aligned}$$

as claimed. □

A.3 Proof of Theorem 2.4.4

We shall derive a relationship using entropy at $s_0 \in (0,\infty)$, then take the limit as $s_0 \to \infty$ to obtain the result. Using Lemma A.2.1(i) we can write

$$\begin{aligned} I(H;\gamma;s_0) &= \frac{1}{2\pi}\int_{-\infty}^{\infty} \mathrm{trace}[H^*(j\omega)H(j\omega)]\left[\frac{s_0^2}{s_0^2+\omega^2}\right] d\omega + O(\gamma^{-2}) \\ &= \frac{1}{2\pi}\int_{-\infty}^{\infty} \mathrm{trace}\left\{\left[H(j\omega)\frac{s_0}{(s_0+j\omega)}\right]^* \right. \\ &\qquad \left. \times \left[H(j\omega)\frac{s_0}{(s_0+j\omega)}\right]\right\} d\omega + O(\gamma^{-2}). \end{aligned}$$

Therefore,

$$I(H;\gamma;s_0) = \left\| \left[\frac{s_0}{s+s_0}\right] H(s) \right\|_2^2 + O(\gamma^{-2}). \tag{A.1}$$

The integrands in (A.1) are monotonically increasing with s_0, and are continuous. Hence, by dominated convergence, both sides of (A.1) tend to a limit as $s_0 \to \infty$. Each side is finite because $H(\infty) = 0$ by assumption (see Proposition 2.3.1(iii) and (iv) and Remark 2.4.3). So, taking $s_0 \to \infty$ and noting that the $O(\gamma^{-2})$ term may be bounded independently of s_0,

$$I(H;\gamma;\infty) = \|H\|_2^2 + O(\gamma^{-2}), \tag{A.2}$$

which proves Part (ii). Part (i) of the theorem follows from this on noting (using Lemma A.2.1(ii)) that the $O(\gamma^{-2})$ terms are non-negative. Finally, Part (iii) is obtained by taking $\gamma \to \infty$ in (A.2). □

A.4 State-Space Evaluation of the Entropy Integral

The following lemmas are useful in establishing state-space formulae for entropy integrals. In particular, Lemma A.4.2 is used in the derivation of the minimum value of the entropy in Theorem 3.5.1. Before stating and proving Lemma A.4.2, it is convenient to state and prove a more restricted result, which we will also need in Section 5.3 and in the proof of the entropy formula in Theorem 4.5.2.

Lemma A.4.1 *Suppose* $G = \left[\begin{array}{c|c} \bar{A} & \bar{B} \\ \hline \bar{C} & \bar{D} \end{array}\right]$ *is a square transfer function matrix with* $\det \bar{D} = 1$ *and* $G^{\pm 1} \in \mathcal{RH}_\infty$. *Then*

$$\lim_{s_0\to\infty} \left\{ -\frac{1}{2\pi} \int_{-\infty}^{\infty} \ln|\det G^*(j\omega)G(j\omega)| \left[\frac{s_0^2}{s_0^2+\omega^2}\right] d\omega \right\} = -\operatorname{trace}[\bar{D}^{-1}\bar{C}\bar{B}].$$

Proof Using the fact that $\ln|\det G^*(j\omega)G(j\omega)| = 2\ln|\det G(j\omega)|$ we can rewrite the integral as

$$I_1 := \lim_{s_0\to\infty} \left\{ -\frac{1}{\pi} \int_{-\infty}^{\infty} \ln|\det G(j\omega)| \left[\frac{s_0^2}{s_0^2+\omega^2}\right] d\omega \right\}.$$

By assumption, $G^{\pm 1} \in \mathcal{RH}_\infty$, which permits the use of Poisson's Integral Theorem [52, Theorem 17.16]:

$$\begin{aligned} I_1 &= -\lim_{s_0\to\infty} \{s_0 \ln|\det G(s_0)|\} \\ &= -\lim_{s_0\to\infty} \{s_0 \ln|\det(\bar{D} + \bar{C}(s_0 I - \bar{A})^{-1}\bar{B})|\} \\ &= -\lim_{s_0\to\infty} \{s_0 \ln|\det \bar{D}| + s_0 \ln|\det(I + \bar{D}^{-1}\bar{C}(s_0 I - \bar{A})^{-1}\bar{B})|\}. \end{aligned}$$

The first term in this last expression is zero because $\det \bar{D} = 1$ by assumption. A power series expansion of the second term in terms of s_0^{-1} gives

$$\begin{aligned} I_1 &= -\lim_{s_0\to\infty}\{s_0 \ln|\det(I+\bar{D}^{-1}\bar{C}(s_0I-\bar{A})^{-1}\bar{B})|\} \\ &= -\lim_{s_0\to\infty}\{s_0 \ln|\det(I+s_0^{-1}\bar{D}^{-1}\bar{C}\bar{B}+O(s_0^{-2}))|\} \\ &= -\lim_{s_0\to\infty}\{\operatorname{trace}[\bar{D}^{-1}\bar{C}\bar{B}]+O(s_0^{-1})\} \\ &= -\operatorname{trace}[\bar{D}^{-1}\bar{C}\bar{B}] \end{aligned}$$

as required, having used Lemma A.2.1(i) to obtain the third equality. □

By relaxing the assumption that $G^{-1} \in \mathcal{RH}_\infty$ we obtain the following lemma, which is the one we need in the proof of the entropy formula of Theorem 3.5.1. A similar result for scalar systems has been derived independently in a different context in [1].

Lemma A.4.2 *Suppose* $G = \left[\begin{array}{c|c} \bar{A} & \bar{B} \\ \hline \bar{C} & \bar{D} \end{array}\right]$ *is a square transfer function matrix with* $\det \bar{D} = 1$ *and* $G \in \mathcal{RH}_\infty$*, but with no assumptions on the stability of* G^{-1}*. Then*

$$\lim_{s_0\to\infty}\left\{-\frac{1}{2\pi}\int_{-\infty}^{\infty}\ln|\det G^*(j\omega)G(j\omega)|\left[\frac{s_0^2}{s_0^2+\omega^2}\right]d\omega\right\} = -\operatorname{trace}[\bar{D}^{-1}\bar{C}\bar{B}] - 2\sum_{\operatorname{Re}\{\lambda_i\}>0}\lambda_i\{\bar{A}-\bar{B}\bar{D}^{-1}\bar{C}\}.$$

Proof It is convenient to define

$$\begin{aligned} F := \left[\begin{array}{c|c} A & B \\ \hline C & D \end{array}\right] &:= G^{-1} \\ &= \left[\begin{array}{c|c} \bar{A}-\bar{B}\bar{D}^{-1}\bar{C} & \bar{B}\bar{D}^{-1} \\ \hline -\bar{D}^{-1}\bar{C} & \bar{D}^{-1} \end{array}\right]. \end{aligned} \tag{A.3}$$

Clearly,

$$\begin{aligned} I_2 &:= \lim_{s_0\to\infty}\left\{-\frac{1}{2\pi}\int_{-\infty}^{\infty}\ln|\det G^*(j\omega)G(j\omega)|\left[\frac{s_0^2}{s_0^2+\omega^2}\right]d\omega\right\} \\ &= \lim_{s_0\to\infty}\left\{\frac{1}{2\pi}\int_{-\infty}^{\infty}\ln|\det F^*(j\omega)F(j\omega)|\left[\frac{s_0^2}{s_0^2+\omega^2}\right]d\omega\right\} \end{aligned}$$

and $F^{-1} \in \mathcal{RH}_\infty$, but F itself is not necessarily in $\mathcal{RH}_\infty$.

Performing an appropriate state transformation on F allows us to write without loss of generality that

$$F = \left[\begin{array}{cc|c} A_1 & 0 & B_1 \\ A_{21} & A_2 & B_2 \\ \hline C_1 & C_2 & D \end{array}\right]$$

where A_1 is not asymptotically stable and A_2 is asymptotically stable. Note that (A_1, B_1) is completely controllable because $\bar{A}$ is asymptotically stable. Let W_1 be the unique positive definite, symmetric, stabilizing solution to

$$0 = A_1^T W_1 + W_1 A_1 - W_1 B_1 B_1^T W_1 \tag{A.4}$$

and define

$$W := \begin{bmatrix} W_1 & 0 \\ 0 & 0 \end{bmatrix}.$$

Then, because W_1 is the stabilizing solution of (A.4), $A - BB^TW$ is easily seen to be asymptotically stable. In addition, it can be verified that

$$U := \left[\begin{array}{c|c} A - BB^TW & B \\ \hline -B^TW & I \end{array}\right]$$

is all-pass, and that

$$\begin{aligned} \tilde{F} := FU &= \left[\begin{array}{cc|c} A & BB^TW & B \\ 0 & A - BB^TW & B \\ \hline C & -DB^TW & D \end{array}\right] \\ &= \left[\begin{array}{c|c} A - BB^TW & B \\ \hline C - DB^TW & D \end{array}\right] \end{aligned}$$

after applying a state transformation of $\begin{bmatrix} I & -I \\ 0 & I \end{bmatrix}$ and removing uncontrollable states. Now $\tilde{F} \in \mathcal{RH}_\infty$ by construction, and also

$$\tilde{F}^{-1} = \left[\begin{array}{c|c} A - BD^{-1}C & BD^{-1} \\ \hline -D^{-1}C + B^TW & D^{-1} \end{array}\right]$$

is in $\mathcal{RH}_\infty$ because $G \in \mathcal{RH}_\infty$ by assumption. Then, exploiting the all-pass nature of U,

$$\det F^*F = \det FF^* = \det FUU^*F^* = \det \tilde{F}\tilde{F}^* = \det \tilde{F}^*\tilde{F},$$

and hence

$$\begin{aligned} I_2 &= \lim_{s_0 \to \infty} \left\{ \frac{1}{2\pi} \int_{-\infty}^{\infty} \ln|\det F^*(j\omega)F(j\omega)| \left[\frac{s_0^2}{s_0^2 + \omega^2}\right] d\omega \right\} \\ &= \lim_{s_0 \to \infty} \left\{ \frac{1}{2\pi} \int_{-\infty}^{\infty} \ln|\det \tilde{F}^*(j\omega)\tilde{F}(j\omega)| \left[\frac{s_0^2}{s_0^2 + \omega^2}\right] d\omega \right\} \\ &= \operatorname{trace}[D^{-1}(C - DB^TW)B] \\ &= \operatorname{trace}[D^{-1}CB] - \operatorname{trace}[B^TWB], \end{aligned}$$

where the third equality is by Lemma A.4.1.

The second term may be further simplified in the following way:

$$\begin{aligned}\text{trace}[B^TWB] &= \text{trace}[B_1^TW_1B_1]\\ &= \text{trace}[W_1B_1B_1^T]\\ &= \text{trace}[A_1^T + W_1A_1W_1^{-1}]\\ &= \text{trace}[A_1^T + A_1]\\ &= 2\sum_i \lambda_i\{A_1\}\\ &= 2\sum_{\text{Re}\{\lambda_i\}>0} \lambda_i\{A\},\end{aligned}$$

where the third equality is from (A.4)W_1^{-1}. Hence, using the definition of A, B, C and D in (A.3), it follows that

$$I_2 = -\ \text{trace}[\bar{D}^{-1}\bar{C}\bar{B}] - 2\sum_{\text{Re}\{\lambda_i\}>0} \lambda_i\{\bar{A} - \bar{B}\bar{D}^{-1}\bar{C}\}$$

as claimed. □

A.5 Proof of Lemma 8.4.9

To show that $\beta G = \tilde{M}^{-1}\beta\tilde{N}$ is a normalized left-coprime factorization of βG we need to show that:

(i) $\beta\tilde{N} \in \mathcal{RH}_\infty$ and $\tilde{M} \in \mathcal{RH}_\infty$.

(ii) $\tilde{M}^{-1}$ exists.

(iii) $\beta\tilde{N}$ and $\tilde{M}$ are left-coprime.

(iv) $\beta^2\tilde{N}\tilde{N}^* + \tilde{M}\tilde{M}^* = I$.

(v) $\beta G = \tilde{M}^{-1}\beta\tilde{N}$.

Proof *Part (i)* is immediate on noticing that $A - \beta^2 Y_\infty C^T C$ is asymptotically stable because Y_∞ is the stabilizing solution of the HFARE.

Part (ii) is true because $\tilde{M}(\infty) = I$, which is nonsingular.

Part (iii) follows from [45].

Part (iv) is verified by routine state-space calculation, as follows.

$$\begin{aligned}\beta^2\tilde{N}\tilde{N}^* + \tilde{M}\tilde{M}^* &= [\beta\tilde{N}\ \ \tilde{M}][\beta\tilde{N}\ \ \tilde{M}]^*\\ &= \left[\begin{array}{c|cc} A - \beta^2 Y_\infty C^T C & \beta B & -\beta^2 Y_\infty C^T \\ \hline C & 0 & I \end{array}\right]\end{aligned}$$

$$\times\left[\begin{array}{c|c} -(A-\beta^2 Y_\infty C^T C)^T & -C^T \\ \beta B^T & 0 \\ -\beta^2 C Y_\infty & I \end{array}\right]$$

$$= \left[\begin{array}{cc|c} A-\beta^2 Y_\infty C^T C & \beta^2(BB^T+\beta^2 Y_\infty C^T C Y_\infty) & -\beta^2 Y_\infty C^T \\ 0 & -(A-\beta^2 Y_\infty C^T C)^T & -C^T \\ \hline C & -\beta^2 C Y_\infty & I \end{array}\right]$$

Apply a state transformation $\begin{bmatrix} I & -\beta^2 Y_\infty \\ 0 & I \end{bmatrix}$ to get

$$\beta^2 \bar{N}\bar{N}^* + \bar{M}\bar{M}^* = \left[\begin{array}{cc|c} A-\beta^2 Y_\infty C^T C & (*) & 0 \\ 0 & -(A-\beta^2 Y_\infty C^T C)^T & -C^T \\ \hline C & 0 & I \end{array}\right], \tag{A.5}$$

where

$$\begin{aligned}(*) &= \beta^2(BB^T - \beta^2 Y_\infty C^T C Y_\infty + AY_\infty + Y_\infty A^T) \\ &= 0,\end{aligned}$$

from the HFARE. Hence all the states of (A.5) are either unobservable or uncontrollable; these may all be deleted to leave $\beta^2\bar{N}\bar{N}^* + \bar{M}\bar{M}^* = I$.

Part (v) Put $v = \beta\bar{N}u_1 + (\bar{M}-I)u_2$, say. If $u_2 = -v$ then

$$v = \beta\bar{N}u_1 + (I-\bar{M})v \implies \bar{M}v = \beta\bar{N}u_1 \implies v = \bar{M}^{-1}\beta\bar{N}u_1. \tag{A.6}$$

This gives a simple means of forming the state-space realization of $\bar{M}^{-1}\beta\bar{N}$ in terms of the realization of $[\beta\bar{N} \;\; \bar{M}]$. To do this, begin with the state-space equations for $[\beta\bar{N} \;\; (\bar{M}-I)]$ which are:

$$\begin{aligned}\dot{x} &= (A-\beta^2 Y_\infty C^T C)x + \beta B u_1 - \beta^2 Y_\infty C^T u_2, \\ v &= Cx.\end{aligned}$$

Defining $u_2 = -v$ as above, gives

$$\begin{aligned}\dot{x} &= (A-\beta^2 Y_\infty C^T C)x + \beta B u_1 + \beta^2 Y_\infty C^T C x \\ &= Ax + \beta B u_1, \\ v &= Cx.\end{aligned}$$

Hence,

$$\begin{aligned}v &= C(sI-A)^{-1}\beta B u_1 \\ &= \beta G u_1.\end{aligned}$$

The result follows on comparing this with (A.6). □

A.6 Proof of Corollary 8.4.19 and Proposition 8.4.21

Proof of Corollary 8.4.19 We only have to prove that (8.16) holds: the corollary follows immediately from that and Proposition 8.4.18. Now, K_r is the Normalized $\mathcal{H}_\infty$ Controller for the plant G_r: the closed loop transfer function is

$$H_{rr} = \begin{bmatrix} S_{rr}G_r & S_{rr}G_rK_r \\ K_rS_{rr}G_r & K_rS_{rr} \end{bmatrix}$$

where $S_{rr} = (I - G_rK_r)^{-1}$, and we have $\|H_{rr}\|_\infty = \hat{\gamma} < \gamma$. We need to rewrite the closed-loop transfer function in terms of the coprime factorizations $G_r = \tilde{M}_r^{-1}\tilde{N}_r$ and $K_r = U_rV_r^{-1}$. We have

$$R_{rr} = \tilde{M}_rV_r - \tilde{N}_rU_r$$

by definition and

$$S_{rr} = V_rR_{rr}^{-1}\tilde{M}_r$$

from (8.13). Straightforward manipulations then give

$$\begin{aligned} H_{rr} &= \begin{bmatrix} S_{rr}G_r & S_{rr} \\ K_rS_{rr}G_r & K_rS_{rr} \end{bmatrix} - \begin{bmatrix} 0 & I \\ 0 & 0 \end{bmatrix} \\ &= \begin{bmatrix} V_rR_{rr}^{-1}\tilde{N}_r & V_rR_{rr}^{-1}\tilde{M}_r \\ U_rR_{rr}^{-1}\tilde{N}_r & U_rR_{rr}^{-1}\tilde{M}_r \end{bmatrix} - \begin{bmatrix} 0 & I \\ 0 & 0 \end{bmatrix} \\ &= \begin{bmatrix} V_r \\ U_r \end{bmatrix} R_{rr}^{-1} \begin{bmatrix} \tilde{N}_r & \tilde{M}_r \end{bmatrix} - \begin{bmatrix} 0 & I \\ 0 & 0 \end{bmatrix}. \end{aligned}$$

Hence

$$\begin{bmatrix} V_r \\ U_r \end{bmatrix} R_{rr}^{-1} \begin{bmatrix} \tilde{N}_r & \tilde{M}_r \end{bmatrix} = \begin{bmatrix} 0 & I \\ 0 & 0 \end{bmatrix} + H_{rr} \tag{A.7}$$

and

$$\begin{bmatrix} \beta V_r \\ U_r \end{bmatrix} R_{rr}^{-1} \begin{bmatrix} \tilde{N}_r & \tilde{M}_r \end{bmatrix} = \begin{bmatrix} \beta I & 0 \\ 0 & I \end{bmatrix} \left(\begin{bmatrix} 0 & I \\ 0 & 0 \end{bmatrix} + H_{rr} \right). \tag{A.8}$$

Similarly,

$$\begin{bmatrix} V_r \\ U_r \end{bmatrix} R_{rr}^{-1} \begin{bmatrix} \beta\tilde{N}_r & \tilde{M}_r \end{bmatrix} = \left(\begin{bmatrix} 0 & I \\ 0 & 0 \end{bmatrix} + H_{rr} \right) \begin{bmatrix} \beta I & 0 \\ 0 & I \end{bmatrix}, \tag{A.9}$$

and also

$$\begin{aligned} \begin{bmatrix} \beta V_r \\ U_r \end{bmatrix} R_{rr}^{-1} \begin{bmatrix} \beta\tilde{N}_r & \tilde{M}_r \end{bmatrix} &= \begin{bmatrix} \beta I & 0 \\ 0 & I \end{bmatrix} \left(\begin{bmatrix} 0 & I \\ 0 & 0 \end{bmatrix} + H_{rr} \right) \begin{bmatrix} \beta I & 0 \\ 0 & I \end{bmatrix} \\ &= \begin{bmatrix} 0 & \beta I \\ 0 & 0 \end{bmatrix} + \begin{bmatrix} \beta I & 0 \\ 0 & I \end{bmatrix} H_{rr} \begin{bmatrix} \beta I & 0 \\ 0 & I \end{bmatrix}. \end{aligned} \tag{A.10}$$

Take the $\mathcal{H}_\infty$-norm of (A.7)–(A.10) and use the triangle inequality and the sub-multiplicative property of the $\mathcal{H}_\infty$-norm. Also use that $\|H_{rr}\|_\infty = \hat{\gamma}$ by definition and that $0 < \beta \leq 1$. Then

$$\left\| \begin{bmatrix} V_r \\ U_r \end{bmatrix} R_{rr}^{-1} \begin{bmatrix} \bar{N}_r & \bar{M}_r \end{bmatrix} \right\|_\infty \leq 1 + \hat{\gamma} \tag{A.11}$$

and

$$\left\| \begin{bmatrix} \beta V_r \\ U_r \end{bmatrix} R_{rr}^{-1} \begin{bmatrix} \bar{N}_r & \bar{M}_r \end{bmatrix} \right\|_\infty \leq \beta + \hat{\gamma}. \tag{A.12}$$

Also,

$$\left\| \begin{bmatrix} V_r \\ U_r \end{bmatrix} R_{rr}^{-1} \begin{bmatrix} \beta\bar{N}_r & \bar{M}_r \end{bmatrix} \right\|_\infty = \left\| \begin{bmatrix} V_r \\ U_r \end{bmatrix} R_{rr}^{-1} \right\|_\infty \leq 1 + \hat{\gamma}, \tag{A.13}$$

and finally

$$\left\| \begin{bmatrix} \beta V_r \\ U_r \end{bmatrix} R_{rr}^{-1} \begin{bmatrix} \beta\bar{N}_r & \bar{M}_r \end{bmatrix} \right\|_\infty = \left\| \begin{bmatrix} \beta V_r \\ U_r \end{bmatrix} R_{rr}^{-1} \right\|_\infty \leq \beta + \hat{\gamma}. \tag{A.14}$$

The first equality in (A.13) and in (A.14) arises because $[\beta\bar{N}_r \;\; \bar{M}_r]$ is normalized. Equation (A.14) is (8.16) as claimed. The remaining equations (A.11)–(A.13) will be needed later. □

Proof of Proposition 8.4.21 We shall stick to the notation and definitions used in the above proof. Introduce

$$R_r := \bar{M}V_r - \bar{N}U_r,$$

then

$$H_r = \begin{bmatrix} V_r \\ U_r \end{bmatrix} R_r^{-1} \begin{bmatrix} \bar{N} & \bar{M} \end{bmatrix} - \begin{bmatrix} 0 & I \\ 0 & 0 \end{bmatrix}. \tag{A.15}$$

Noting that

$$R_r = R_{rr} + (\beta^{-1}\Delta_{\bar{M}})(\beta V_r) - \Delta_{\bar{N}}U_r = R_{rr} - \begin{bmatrix} \Delta_{\bar{N}} & -\beta^{-1}\Delta_{\bar{M}} \end{bmatrix} \begin{bmatrix} U_r \\ \beta V_r \end{bmatrix},$$

an application of the well-known matrix inversion lemma [32, p19] gives

$$R_r^{-1} = R_{rr}^{-1} + R_{rr}^{-1} \begin{bmatrix} \Delta_{\bar{N}} & -\beta^{-1}\Delta_{\bar{M}} \end{bmatrix} T^{-1} \begin{bmatrix} U_r \\ \beta V_r \end{bmatrix} R_{rr}^{-1}, \tag{A.16}$$

where

$$T := I - \begin{bmatrix} U_r \\ \beta V_r \end{bmatrix} R_{rr}^{-1} \begin{bmatrix} \Delta_{\bar{N}} & -\beta^{-1}\Delta_{\bar{M}} \end{bmatrix}.$$

Substitute (A.16) into the expression for H_r in (A.15) and use the identity $[\bar{N} \;\; \bar{M}] = [\bar{N}_r + \Delta_{\bar{N}} \;\; \bar{M}_r + \Delta_{\bar{M}}]$ to obtain

$$\begin{aligned} H_r = H_{rr} &+ \begin{bmatrix} V_r \\ U_r \end{bmatrix} R_{rr}^{-1} \begin{bmatrix} \Delta_{\bar{N}} & \beta^{-1}\Delta_{\bar{M}} \end{bmatrix} \begin{bmatrix} I & 0 \\ 0 & \beta I \end{bmatrix} \\ &+ \begin{bmatrix} V_r \\ U_r \end{bmatrix} R_{rr}^{-1} \begin{bmatrix} \Delta_{\bar{N}} & -\beta^{-1}\Delta_{\bar{M}} \end{bmatrix} \\ &\qquad \times T^{-1} \begin{bmatrix} 0 & I \\ I & 0 \end{bmatrix} \begin{bmatrix} \beta V_r \\ U_r \end{bmatrix} R_{rr}^{-1} \begin{bmatrix} \bar{N}_r & \bar{M}_r \end{bmatrix} \\ &+ \begin{bmatrix} V_r \\ U_r \end{bmatrix} R_{rr}^{-1} \begin{bmatrix} \Delta_{\bar{N}} & -\beta^{-1}\Delta_{\bar{M}} \end{bmatrix} \\ &\qquad \times T^{-1} \begin{bmatrix} 0 & I \\ I & 0 \end{bmatrix} \begin{bmatrix} \beta V_r \\ U_r \end{bmatrix} R_{rr}^{-1} \begin{bmatrix} \Delta_{\bar{N}} & \beta^{-1}\Delta_{\bar{M}} \end{bmatrix} \begin{bmatrix} I & 0 \\ 0 & \beta I \end{bmatrix} \end{aligned}$$

Taking $\mathcal{H}_\infty$-norms, and using the triangle inequality and sub-multiplicative property of the $\mathcal{H}_\infty$-norm, leads to

$$\begin{aligned} \|H_r\|_\infty \leq \|H_{rr}\|_\infty &+ \left\| \begin{bmatrix} V_r \\ U_r \end{bmatrix} R_{rr}^{-1} \right\|_\infty \| \begin{bmatrix} \Delta_{\bar{N}} & \beta^{-1}\Delta_{\bar{M}} \end{bmatrix} \|_\infty \\ &+ \left\| \begin{bmatrix} V_r \\ U_r \end{bmatrix} R_{rr}^{-1} \right\|_\infty \left\| \begin{bmatrix} \beta V_r \\ U_r \end{bmatrix} R_{rr}^{-1} \begin{bmatrix} \bar{N}_r & \bar{M}_r \end{bmatrix} \right\|_\infty \\ &\qquad \times \| \begin{bmatrix} \Delta_{\bar{N}} & \beta^{-1}\Delta_{\bar{M}} \end{bmatrix} \|_\infty \|T^{-1}\|_\infty \\ &+ \left\| \begin{bmatrix} V_r \\ U_r \end{bmatrix} R_{rr}^{-1} \right\|_\infty \left\| \begin{bmatrix} \beta V_r \\ U_r \end{bmatrix} R_{rr}^{-1} \right\|_\infty \| \begin{bmatrix} \Delta_{\bar{N}} & \beta^{-1}\Delta_{\bar{M}} \end{bmatrix} \|_\infty^2 \|T^{-1}\|_\infty . \end{aligned}$$

To complete the derivation, use (A.11)–(A.14), and recall that $\|H_{rr}\|_\infty = \hat{\gamma}$ by definition and that

$$\| \begin{bmatrix} \Delta_{\bar{N}} & \beta^{-1}\Delta_{\bar{M}} \end{bmatrix} \|_\infty = \hat{\epsilon} \tag{A.17}$$

by definition. Also, using the inequality [32, p301]

$$\|(I-A)^{-1}\|_\infty \leq (1 - \|A\|_\infty)^{-1} \qquad \text{for} \quad \|A\|_\infty < 1, \tag{A.18}$$

we get

$$\begin{aligned} \|T^{-1}\|_\infty &\leq \left(1 - \left\| \begin{bmatrix} U_r \\ \beta V_r \end{bmatrix} R_{rr}^{-1} \begin{bmatrix} \Delta_{\bar{N}} & -\beta^{-1}\Delta_{\bar{M}} \end{bmatrix} \right\|_\infty \right)^{-1} \\ &\leq \left(1 - \left\| \begin{bmatrix} \beta V_r \\ U_r \end{bmatrix} R_{rr}^{-1} \right\|_\infty \| \begin{bmatrix} \Delta_{\bar{N}} & \beta^{-1}\Delta_{\bar{M}} \end{bmatrix} \|_\infty \right)^{-1} \\ &\leq (1 - (\beta + \hat{\gamma})\hat{\epsilon})^{-1}, \end{aligned}$$

where the last inequality follows from (A.14) and (A.17). (Validity of the use of (A.18) and the existence of the necessary inverses is assured because $(\beta+\hat{\gamma})\hat{\epsilon} < 1$ by assumption.) Collecting all of the above together, gives

$$\begin{aligned}\|H_r\|_\infty &\le \hat{\gamma} + (1+\hat{\gamma})\hat{\epsilon} + (1+\hat{\gamma})(\beta+\hat{\gamma})(1-(\beta+\hat{\gamma})\hat{\epsilon})^{-1}\hat{\epsilon} \\ &\quad + (1+\hat{\gamma})(\beta+\hat{\gamma})(1-(\beta+\hat{\gamma})\hat{\epsilon})^{-1}\hat{\epsilon}^2 \\ &= \hat{\gamma} + \frac{\hat{\epsilon}(1+\hat{\gamma})(1+\beta+\hat{\gamma})}{(1-\hat{\epsilon}(\beta+\hat{\gamma}))},\end{aligned}$$

as claimed. □

A.7 Proof of Theorem 9.2.2(ii)

Differentiate $Z = \gamma^{-2}XY - I$ to obtain

$$\frac{dZ}{d\gamma} = -2\gamma^{-1}(I+Z) + \gamma^{-2}\frac{dX}{d\gamma}Y + \gamma^{-2}X\frac{dY}{d\gamma}.$$

(Existence of the derivatives is assured by [50].) By differentiating $(Z^{-1})Z = I$ we get

$$\begin{aligned}&\frac{d}{d\gamma}(Z^{-1}) = -Z^{-1}\frac{dZ}{d\gamma}Z^{-1} \\ \Longrightarrow\quad &\frac{d}{d\gamma}(Z^{-1}) = 2\gamma^{-1}Z^{-1} + 2\gamma^{-1}Z^{-2} - \gamma^{-2}Z^{-1}\frac{dX}{d\gamma}YZ^{-1} - \gamma^{-2}Z^{-1}X\frac{dY}{d\gamma}Z^{-1}\end{aligned}$$

which will be of use later.

The expressions for the derivatives of $\hat{A}$, $\hat{B}$ and $\hat{C}$ are obtained by straightforward manipulations of the state-space formulae after using the product rule for differentiation, and substituting for $d(Z^{-1})/d\gamma$ when necessary. Firstly,

$$\frac{d\hat{C}}{d\gamma} = \frac{d}{d\gamma}(-\gamma^{-1}C_2X) = \gamma^{-2}C_2X - \gamma^{-1}C_2\frac{dX}{d\gamma}.$$

Then

$$\begin{aligned}\frac{d\hat{B}}{d\gamma} &= \frac{d}{d\gamma}(\gamma^{-1}YZ^{-1}B_2) \\ &= -\gamma^{-2}YZ^{-1}B_2 + \gamma^{-1}\frac{dY}{d\gamma}Z^{-1}B_2 \\ &\quad + \gamma^{-1}Y\left(2\gamma^{-1}Z^{-1} + 2\gamma^{-1}Z^{-2} - \gamma^{-2}Z^{-1}\frac{dX}{d\gamma}YZ^{-1} - \gamma^{-2}Z^{-1}X\frac{dY}{d\gamma}Z^{-1}\right)B_2 \\ &= \gamma^{-2}YZ^{-1}B_2 - \gamma^{-1}(\gamma^{-2}YZ^{-1}X - I)\frac{dY}{d\gamma}Z^{-1}B_2\end{aligned}$$

$$
\begin{aligned}
&+ 2\gamma^{-2} Y Z^{-2} B_2 - \gamma^{-3} Y Z^{-1} \frac{dX}{d\gamma} Y Z^{-1} B_2 \\
&= \gamma^{-2} Y Z^{-1} B_2 - \gamma^{-1} Z^{-T} \frac{dY}{d\gamma} Z^{-1} B_2 \\
&\quad + 2\gamma^{-2} Z^{-T} Y Z^{-1} B_2 - \gamma^{-3} Y Z^{-1} \frac{dX}{d\gamma} Y Z^{-1} B_2 \\
&= \gamma^{-1} Z^{-T} \bar{Y} Z^{-1} B_2 + \gamma^{-2} Y Z^{-1} B_2 - \gamma^{-3} Y Z^{-1} \frac{dX}{d\gamma} Y Z^{-1} B_2.
\end{aligned}
$$

Finally,

$$
\begin{aligned}
\frac{d\hat{A}}{d\gamma} &= \frac{d}{d\gamma}(-A^T - \gamma^{-1} \hat{B} B_2^T - \gamma^{-2} C_1^T C_1 X) \\
&= \gamma^{-2} \hat{B} B_2^T - \gamma^{-1} \frac{d\hat{B}}{d\gamma} B_2^T + 2\gamma^{-3} C_1^T C_1 X - \gamma^{-2} C_1^T C_1 \frac{dX}{d\gamma} \\
&= \gamma^{-3} Y Z^{-1} B_2 B_2^T - \gamma^{-3} Y Z^{-1} B_2 B_2^T - \gamma^{-2} Z^{-T} \bar{Y} Z^{-1} B_2 B_2^T \\
&\quad + \gamma^{-4} Y Z^{-1} \frac{dX}{d\gamma} Y Z^{-1} B_2 B_2^T + \gamma^{-2} C_1^T C_1 \bar{X} \\
&= -\gamma^{-2} Z^{-T} \bar{Y} Z^{-1} B_2 B_2^T + \gamma^{-2} C_1^T C_1 \bar{X} + \gamma^{-4} Y Z^{-1} \frac{dX}{d\gamma} Y Z^{-1} B_2 B_2^T,
\end{aligned}
$$

which completes the proof. □

Appendix B

Entropy Formulae: Alternative Derivation

B.1 Introduction

Our purpose here is to provide an alternative derivation of the entropy formulae of Section 3.5, in the spirit of the recent work [17, 28] and its extensions in [29]. Done this way, the derivation is less manipulative and more intuitive. Proof of Theorem 3.5.1 is the main aim; proof of Lemma 3.5.5 is also included.

Assumptions and notation is inherited from Chapter 3 and Section 3.5. In particular, P is a standard plant satisfying the Assumptions 3.2.1 and 3.2.7; also $D_{11} = 0$. In keeping with [17, 28, 29] we assume throughout this appendix that $\gamma = 1$. This results in no loss of generality since it is achieved by the scalings $\gamma^{-1/2}B_1$, $\gamma^{-1/2}C_1$, $\gamma^{1/2}B_2$, $\gamma^{1/2}C_2$, and $\gamma^{-1}K$. We will therefore talk of $(P, 1)$-admissible controllers and closed-loops rather than (P, γ)-admissible controllers and closed-loops. Note that there are some notational differences between this monograph and [17, 28, 29].

B.2 The Full Information and Output Estimation Problems

In this section we are concerned with the *Full Information* and *Output Estimation* problems. These problems are used in [17, 29] as stepping stones to the parametrization of all $(P, 1)$-admissible controllers.

The first lemma deals with the Full Information problem, where the controller has direct access to both the state and the disturbance input.

Lemma B.2.1 (Full Information [29, Theorem 3.1]) *Consider the 'Full Information' standard plant*

$$P_{FI} = \left[\begin{array}{c|cc} A & B_1 & B_2 \\ \hline C_1 & 0 & D_{12} \\ \begin{bmatrix} I \\ 0 \end{bmatrix} & \begin{bmatrix} 0 \\ I \end{bmatrix} & \begin{bmatrix} 0 \\ 0 \end{bmatrix} \end{array}\right]$$

satisfying Assumptions 3.2.1 and 3.2.7 as appropriate. Let $D_\perp$ make $[D_{12} \;\; D_\perp]$ square and orthogonal. Then

(i) *There exists a $(P_{FI}, 1)$-admissible controller if and only if there exists $X_\infty = X_\infty^T \geq 0$ solving the Full Information algebraic Riccati equation*

$$\begin{aligned} 0 = X_\infty(A - B_2 D_{12}^T C_1) + (A - B_2 D_{12}^T C_1)^T X_\infty \\ + C_1^T D_\perp D_\perp^T C_1 + X_\infty(B_1 B_1^T - B_2 B_2^T) X_\infty \end{aligned}$$

such that

$$A - B_2 D_{12}^T C_1 + (B_1 B_1^T - B_2 B_2^T) X_\infty \quad \textit{is asymptotically stable.}$$

(ii) *All $(P_{FI}, 1)$-admissible closed-loops H_{FI} are generated by*

$$H_{FI} = \mathcal{F}(R, \Phi) \qquad \text{where } \Phi \in \mathcal{BRH}_\infty,$$

where

$$R = \begin{bmatrix} R_{11} & R_{12} \\ R_{21} & R_{22} \end{bmatrix} := \left[\begin{array}{c|cc} A_F & B_1 & B_2 \\ \hline C_{1F} & 0 & D_{12} \\ -B_1^T X_\infty & I & 0 \end{array}\right]$$

with

$$A_F := A - B_2(D_{12}^T C_1 + B_2^T X_\infty)$$
$$C_{1F} := C_1 - D_{12}(D_{12}^T C_1 + B_2^T X_\infty).$$

(iii) *The transfer function matrix R given in (ii) above has the following properties: $R \in \mathcal{RH}_\infty$, $R^* R = I$ and $R_{21}^{\pm 1} \in \mathcal{RH}_\infty$.*

In the general output feedback problem, if $D_{21} = I$ and $A - B_1 C_2$ is asymptotically stable then the state and the disturbance input can be reconstructed exactly by using an observer. Hence the achievable closed-loops would be identical to those of the Full Information problem. Taking the transpose of the above result leads to the following corollary for the Output Estimation problem.

Corollary B.2.2 (Output Estimation [29]) *Consider the 'Output Estimation' standard plant*

$$P_{OE} = \left[\begin{array}{c|cc} A & B_1 & B_2 \\ \hline C_1 & 0 & I \\ C_2 & D_{21} & 0 \end{array}\right]$$

satisfying Assumptions 3.2.1 and 3.2.7 as appropriate, together with $A - B_2 C_1$ asymptotically stable. Let $\tilde{D}_\perp$ make $[D_{21}^T \;\; \tilde{D}_\perp^T]^T$ square and orthogonal. Then

(i) *There exists a $(P_{OE}, 1)$-admissible controller if and only if there exists $Y_\infty = Y_\infty^T \geq 0$ solving the Output Estimation algebraic Riccati equation*

$$0 = Y_\infty(A - B_1 D_{21}^T C_2)^T + (A - B_1 D_{21}^T C_2) Y_\infty + B_1 \tilde{D}_\perp^T \tilde{D}_\perp B_1^T + Y_\infty(\gamma^{-2} C_1^T C_1 - C_2^T C_2) Y_\infty$$

such that

$$A - B_1 D_{21}^T C_2 + Y_\infty(\gamma^{-2} C_1^T C_1 - C_2^T C_2) \quad \text{is asymptotically stable.}$$

(ii) *All $(P_{OE}, 1)$-admissible closed-loops H_{OE} are generated by*

$$H_{OE} = \mathcal{F}(S, \Phi) \qquad \text{where } \Phi \in \mathcal{BRH}_\infty,$$

where

$$S = \begin{bmatrix} S_{11} & S_{12} \\ S_{21} & S_{22} \end{bmatrix} := \left[\begin{array}{c|cc} A_L & B_{1L} & -Y_\infty C_1^T \\ \hline C_1 & 0 & I \\ C_2 & D_{21} & 0 \end{array}\right]$$

with

$$\begin{aligned} A_L &:= A - (B_1 D_{21}^T + Y_\infty C_2^T) C_2 \\ B_{1L} &:= B_1 - (B_1 D_{21}^T + Y_\infty C_2^T) D_{21}. \end{aligned}$$

(iii) *The transfer function matrix S given in (ii) above has the following properties: $S \in \mathcal{RH}_\infty$, $SS^* = I$ and $S_{12}^{\pm 1} \in \mathcal{RH}_\infty$.*

B.3 Separation Structure

Consider the problem of characterizing all $(P, 1)$-admissible controllers K. A key step in the solution of this problem as given by [17, 29] is to separate the problem into a Full Information problem and an Output Estimation problem. This approach will be useful for our purposes, so it will be outlined here.

Define a new standard plant $\bar{P}$, in terms of the given standard plant P, as follows (as in equation (4.7) of [29])

$$\bar{P} = \left[\begin{array}{c|cc} \bar{A} & \bar{B}_1 & \bar{B}_2 \\ \hline \bar{C}_1 & 0 & \bar{D}_{12} \\ \bar{C}_2 & \bar{D}_{21} & 0 \end{array}\right] := \left[\begin{array}{c|cc} A + B_1 B_1^T X_\infty & B_1 & B_2 \\ \hline D_{12}^T C_1 + B_2^T X_\infty & 0 & I \\ C_2 + D_{21} B_1^T X_\infty & D_{21} & 0 \end{array}\right]. \tag{B.1}$$

Then [29, Lemma 4.4] says that a controller K is $(P, 1)$-admissible if and only if it is $(\bar{P}, 1)$-admissible. So we can shift our attention from P onto $\bar{P}$; moreover, $\bar{P}$ satisfies the assumptions for the Output Estimation problem. Furthermore, the feedback connection of K with P gives the same closed-loop transfer function as that given by the feedback connection of K with R with $\bar{P}$, where R is as defined in Lemma B.2.1 for the Full Information problem. So

$$\mathcal{F}(P, K) = \mathcal{F}(R, \mathcal{F}(\bar{P}, K)). \tag{B.2}$$

But now we can apply Corollary B.2.2 to $\bar{P}$: all $(\bar{P}, 1)$-admissible closed-loops are generated by

$$\mathcal{F}(\bar{P}, K) = \mathcal{F}(\bar{S}, \Phi) \quad \text{where } \Phi \in \mathcal{BRH}_\infty, \tag{B.3}$$

with an obvious notation.

Combining (B.2) and (B.3), it follows that all $(P, 1)$-admissible closed-loops H are generated by

$$H = \mathcal{F}(R, \mathcal{F}(\bar{S}, \Phi)) \quad \text{where } \Phi \in \mathcal{BRH}_\infty. \tag{B.4}$$

B.4 Proof of Theorem 3.5.1

The following lemma shows how the separation structure outlined in the previous section manifests itself in the entropy formulae.

Lemma B.4.1 *Let R be as defined in Lemma B.2.1 for the Full Information problem associated with the given standard plant P. Let $\bar{P}$ be as defined in (B.1). Let $\bar{S}$ be as defined in Corollary B.2.2 for the Output Estimation problem based on $\bar{P}$. Let $H_{ME\infty}$ be the minimum entropy $(P,1)$-admissible closed-loop. Then*

$$I(H_{ME\infty};1;\infty) = I(R_{11};1;\infty) + I(\bar{S}_{11};1;\infty).$$

Proof From (B.4), all $(P,1)$-admissible closed-loops H are given by

$$H = \mathcal{F}(R, \mathcal{F}(\bar{S}, \Phi)) \quad \text{where } \Phi \in \mathcal{BRH}_\infty.$$

Setting $\Phi = \Phi_{ME\infty} = 0$ (Theorem 3.4.2) gives

$$H_{ME\infty} = \mathcal{F}(R, \bar{S}_{11}).$$

Since $R^* R = I$, it is easy to verify that

$$I - H_{ME\infty}^* H_{ME\infty} = R_{21}^*(I - R_{22}\bar{S}_{11})^{-*}(I - \bar{S}_{11}^*\bar{S}_{11})(I - R_{22}\bar{S}_{11})^{-1}R_{21}.$$

Substitution into the definition of $I(H_{ME\infty};1;\infty)$ gives

$$I(H_{ME\infty};1;\infty) = I(R_{11};1;\infty) + I(\bar{S}_{11};1;\infty) + a_\infty,$$

where

$$a_\infty := \lim_{s_0 \to \infty} \{a_{s_0}\}$$

and

$$a_{s_0} := \frac{1}{\pi}\int_{-\infty}^{\infty} \ln|\det(I - R_{22}(j\omega)\bar{S}_{11}(j\omega))| \left[\frac{s_0^2}{s_0^2 + \omega^2}\right] d\omega.$$

Now observe that R_{22} and $\bar{S}_{11}$ are strictly proper, so $s_0 R_{22}(s_0)\bar{S}_{11}(s_0) \to 0$ as $s_0 \to \infty$. Also note that $(I - R_{22}\bar{S}_{11})^{\pm 1} \in \mathcal{RH}_\infty$. An application of Lemma A.4.1 then shows that in fact $a_\infty = 0$. □

The above lemma expresses the total entropy as the sum of the 'Full Information entropy' $I(R_{11};1;\infty)$ and the 'Output Estimation entropy' $I(\bar{S}_{11};1;\infty)$. The former is easily evaluated using Lemma A.4.1 since $I - R_{11}^* R_{11} = R_{21}^* R_{21}$ and $R_{21}^{\pm 1} \in \mathcal{RH}_\infty$. We find

$$I(R_{11};1;\infty) = \text{trace}[B_1^T X_\infty B_1].$$

Similarly, the latter entropy is

$$I(\bar{S}_{11};1;\infty) = \text{trace}[\bar{C}_1 \bar{Y}_\infty \bar{C}_1^T].$$

Here $\bar{Y}_\infty$ solves the Output Estimation algebraic Riccati equation associated with $\bar{P}$; this is shown in the proof of Theorem 4.1 of [29] to be

$$\bar{Y}_\infty = Z_\infty Y_\infty. \tag{B.5}$$

To complete the proof, one only has to recall that $\bar{C}_1 = D_{12}^T C_1 + B_2^T X_\infty$. Lemma B.4.1 together with the above expressions then gives

$$\begin{aligned} I(H_{ME\infty}; 1; \infty) = \; & \mathrm{trace}[B_1^T X_\infty B_1] \\ & + \mathrm{trace}[(D_{12}^T C_1 + B_2^T X_\infty) Z_\infty Y_\infty (D_{12}^T C_1 + B_2^T X_\infty)^T], \end{aligned}$$

which is the second entropy formula of Theorem 3.5.1. The other formula given in Theorem 3.5.1 follows by duality. □

B.5 Proof of Lemma 3.5.5

Complete derivations of the characterization of all $(P,1)$-admissible controllers, as stated without proof in [28], have now appeared in [29] using techniques established in [17] (see also [30]). These papers give the characterization of all $(P,1)$-admissible closed-loops but do not give the state-space formulae for the all-pass dilation that is required in Lemma 3.5.5. We fill in here the steps required for that calculation.

As in the previous sections, we exploit the separation structure as in [17, 29] with the assumption that $D_{11} = 0$. There is no difficulty in principle with the $D_{11} \neq 0$; the expressions just become more involved.

Recall the discussion of the separation structure given in Section B.3. We saw in equation (B.4) that all $(P,1)$-admissible closed-loops are generated by

$$H = \mathcal{F}(R, \mathcal{F}(\bar{S}, \Phi)) \quad \text{where } \Phi \in \mathcal{BRH}_\infty.$$

To obtain the all-pass dilation needed in Lemma 3.5.5 we will first dilate R into an all-pass transfer function matrix R_a, then we will dilate $\bar{S}$ into an all-pass transfer matrix $\bar{S}_a$. Augmenting Φ with zero rows and columns as appropriate, after combining R_a and $\bar{S}_a$ into a single feedback element, will give the desired result.

Dilating R into an all-pass transfer function matrix R_a gives

$$R_a = \left[\begin{array}{c|ccc} A_F & B_1 & B_2 & B_3 \\ \hline C_{1F} & 0 & D_{12} & D_\perp \\ -B_1^T X_\infty & I & 0 & 0 \end{array}\right].$$

The observability Gramian of R_a is X_∞ (see the proof of Lemma 3.4 of [29]). It follows from [26, Theorem 5.1] that R_a is all-pass if

$$\begin{bmatrix} 0 & D_{12} & D_\perp \\ I & 0 & 0 \end{bmatrix}^T \begin{bmatrix} C_{1F} \\ -B_1^T X_\infty \end{bmatrix} + [\; B_1 \quad B_2 \quad B_3 \;]^T X_\infty = 0,$$

which is satisfied by

$$B_3 = -X_\infty^{\dagger} C_1^T D_\perp.$$

Before dilating $\bar{S}$, apply a state transformation of $-Z_\infty^{-1}$ to $\bar{P}$ so that

$$\bar{P} = \left[\begin{array}{c|cc} Z_\infty^{-1}(A + B_1 B_1^T X_\infty)Z_\infty & -Z_\infty^{-1}B_1 & -Z_\infty^{-1}B_2 \\ \hline -(D_{12}^T C_1 + B_2^T X_\infty)Z_\infty & 0 & I \\ -(C_2 + D_{21}B_1^T X_\infty)Z_\infty & D_{21} & 0 \end{array}\right]$$

$$= \left[\begin{array}{c|cc} Z_\infty^{-1}(A + B_1 B_1^T X_\infty)Z_\infty & -Z_\infty^{-1}B_1 & -Z_\infty^{-1}B_2 \\ \hline \hat{C}_1 & 0 & I \\ \hat{C}_2 & D_{21} & 0 \end{array}\right].$$

Then $\bar{S}$ is defined in Corollary B.2.2 based on the above $\bar{P}$:

$$\bar{S} = \left[\begin{array}{c|cc} \bar{A}_L & \bar{B}_{1L} & \bar{B}_{2L} \\ \hline \hat{C}_1 & 0 & I \\ \hat{C}_2 & D_{21} & 0 \end{array}\right].$$

Before writing down expressions for $\bar{A}_L$, $\bar{B}_{1L}$ and $\bar{B}_{2L}$ we note that the solution to the Output Estimation algebraic Riccati equation associated with the above $\bar{P}$ is $Y_\infty Z_\infty^{-T} = Z_\infty^{-1} Y_\infty$. (This comes from (B.5) after taking the $-Z_\infty^{-1}$ state transformation into account). Now we can apply Corollary B.2.2 to $\bar{P}$ to get

$$\begin{aligned} \bar{A}_L &= Z_\infty^{-1}(A + B_1 B_1^T X_\infty)Z_\infty - (-Z_\infty^{-1}B_1 D_{21}^T + Y_\infty Z_\infty^{-1}\hat{C}_2^T)\hat{C}_2 \\ &= \hat{A} - \hat{B}_1\hat{C}_2 - Z_\infty^{-1}B_2\hat{C}_1 - ((-I + Y_\infty X_\infty)B_1 D_{21}^T - Y_\infty(C_2^T + X_\infty B_1 D_{21}^T))\hat{C}_2 \\ &= \hat{A} - Z_\infty^{-1}B_2\hat{C}_1, \end{aligned}$$

after substituting from the second expression for $\hat{A}$ and $Z_\infty^{-T}\hat{C}_2^T$ (from the statement of Lemma 3.5.5). Also,

$$\begin{aligned} \bar{B}_{1L} &= -Z_\infty^{-1}B_1 - (-Z_\infty^{-1}B_1 D_{21}^T + Z_\infty^{-1}Y_\infty \hat{C}_2^T)D_{21} \\ &= -Z_\infty^{-1}B_1 + \hat{B}_1 D_{21} \end{aligned}$$

and

$$\begin{aligned} \bar{B}_{2L} &= -Y_\infty Z_\infty^{-T}\hat{C}_1^T \\ &= Y_\infty(C_1^T D_{12} + X_\infty B_2) \\ &= \hat{B}_2 - Z_\infty^{-1}B_2. \end{aligned}$$

Further, the all-pass dilation of $\bar{S}$ is just

$$\bar{S}_a = \left[\begin{array}{c|cc} \bar{A}_L & \bar{B}_{1L} & \bar{B}_{2L} \\ \hline \hat{C}_1 & 0 & I \\ \hat{C}_2 & D_{21} & 0 \\ \hat{C}_3 & \tilde{D}_\perp & 0 \end{array}\right].$$

The feedback combination of R_a with $\bar{S}_a$ will then give an all-pass system as follows

$$R_a: \quad \begin{cases} \dot{x} = A_F x + B_1 w + B_2 v + B_3 w_3 \\ z = C_{1F} x + D_{12} v + D_\perp w_3 \\ r = -B_1^T X_\infty x + w \end{cases}$$

$$\bar{S}_a: \quad \begin{cases} \dot{\hat{x}} = (\hat{A} - Z_\infty^{-1} B_2 \hat{C}_1)\hat{x} + \bar{B}_{1L} r + \bar{B}_{2L} w_2 \\ v = \hat{C}_1 \hat{x} + w_2 \\ z_2 = \hat{C}_2 \hat{x} + D_{21} r \\ z_3 = \hat{C}_3 \hat{x} + \tilde{D}_\perp r \end{cases}$$

Eliminating v and r gives

$$\begin{bmatrix} \dot{x} \\ \dot{\hat{x}} \\ z \\ z_2 \\ z_3 \end{bmatrix} = \left[\begin{array}{cc|ccc} A + B_2\hat{C}_1 Z_\infty^{-1} & B_2\hat{C}_1 & B_1 & B_2 & B_3 \\ -\bar{B}_{1L} B_1^T X_\infty & \hat{A} - Z_\infty^{-1} B_2 \hat{C}_1 & \bar{B}_{1L} & \bar{B}_{2L} & 0 \\ \hline C_{1F} & D_{12}\hat{C}_1 & 0 & D_{12} & D_\perp \\ -D_{21} B_1^T X_\infty & \hat{C}_2 & D_{21} & 0 & 0 \\ -\tilde{D}_\perp B_1^T X_\infty & \hat{C}_3 & \tilde{D}_\perp & 0 & 0 \end{array}\right] \begin{bmatrix} x \\ \hat{x} \\ w \\ w_2 \\ w_3 \end{bmatrix}.$$

Finally a state transformation of $\begin{bmatrix} I & 0 \\ Z_\infty^{-1} & I \end{bmatrix}$ is easily shown to give the realization for J_a given in the statement of the lemma.

The equivalence of the two expressions for $\hat{A}$ given in the lemma is obtained by substitution from the two Riccati equations. □

Appendix C

Notation

C.1 Basic Notational Conventions

All systems are linear, multivariable, finite-dimensional and time-invariant, and possess real-rational transfer function matrices. In general, we use capital letters to represent matrices and lower case letters to represent vectors.

We shall not distinguish between a time domain signal and its Laplace transform. For example, w represents both the time domain signal $w(t)$ and the Laplace domain signal $w(s)$. The context will determine which domain is used. We work exclusively in continuous time. A (proper) transfer function matrix is represented in terms of state-space data by

$$\left[\begin{array}{c|c} A & B \\ \hline C & D \end{array}\right] := D + C(sI - A)^{-1}B$$

or by (A, B, C, D), where A, B, C and D are real matrices of appropriate dimension and I is the identity matrix. If $D = 0$, the zero matrix, then the system is strictly proper and we shall write (A, B, C). The matrix A is *asymptotically stable* if and only if each of its eigenvalues has a strictly negative real part. In that case the system (A, B, C, D) is also called asymptotically stable (or *causal* in Chapter 4). A matrix A is called *antistable* if $-A$ is asymptotically stable. In that case, in Chapter 4, the system (A, B, C, D) is called *anticausal.*

As far as possible standard notation has been used. For convenience some of the global conventions are listed below. Notational conventions defined and used locally in the text in a single section are not listed here. Firstly the spaces of interest—see [24, Chapter 2] for more details.

$\mathcal{RL}_2$ Lebesgue space of real-rational transfer function matrices square-integrable on the imaginary axis.

$\mathcal{RH}_2$ Hardy space of real-rational transfer function matrices square-integrable on the imaginary axis with analytic continuation into the right-half plane.

$\mathcal{RL}_\infty$ Lebesgue space of real-rational transfer function matrices bounded on the imaginary axis.

$\mathcal{RH}_\infty$ Hardy space of real-rational transfer function matrices bounded on the imaginary axis with analytic continuation into the right-half plane.

Next some miscellaneous notation and the definition of the norms on the above spaces.

$\forall$	For each, for all.
$\exists$	There exists.
$X := Y$	X is defined to equal Y.
$X =: Y$	Y is defined to equal X.
$\mathbb{R}$	The real numbers.
$\mathrm{Re}\{(\cdot)\}$	Real part of $(\cdot)$.
$\mathcal{R}$	(Prefix) Real-rational.
$\lambda_i\{M\}$	The ith eigenvalue of a square matrix M.
$\rho(M)$	The spectral radius of a square matrix M.
M^T	The transpose of a matrix M.
M^*	The complex conjugate transpose of a matrix M.
$M^\dagger$	The Moore-Penrose generalized inverse of a matrix M.
$M > 0$	$M = M^*$ is positive definite.
$M \geq 0$	$M = M^*$ is positive semidefinite.
$M^{1/2}$	For a matrix $M \geq 0$, any square matrix L such that $M = L^*L$.
$G^*(s)$	$:= G(-s)^T$, the parahermitian conjugate of $G(s)$.
$\sigma_i\{G(s)\}$	$:= \lambda_i^{1/2}\{G^*(s)G(s)\}$, the ith singular value of $G(s)$; by convention ordered $\sigma_1 \geq \sigma_2 \geq \cdots \geq \sigma_n \geq 0$.
$\Vert G(s)\Vert_2$	$:= \left\{(1/2\pi)\int_{-\infty}^{\infty} \mathrm{trace}[G^*(j\omega)G(j\omega)]d\omega\right\}^{1/2}$, the $\mathcal{L}_2$- and $\mathcal{H}_2$-norm.
$\Vert G(s)\Vert_\infty$	$:= \sup_\omega \sigma_1\{G(j\omega)\}$, the $\mathcal{L}_\infty$- and $\mathcal{H}_\infty$-norm.
$\mathcal{BRH}_\infty$	Open unit ball in $\mathcal{RH}_\infty$ i.e., all $\Phi \in \mathcal{RH}_\infty$ such that $\Vert\Phi\Vert_\infty < 1$.

A square but not necessarily stable transfer function matrix $G(s)$ is said to be *all-pass* if $G^*(s)G(s) = I$ for all s (equivalently, if $G(s)G^*(s) = I$ for all s).

Finally some of the system-theoretic quantities.

$(\cdot)_{ME_{s_0}}$	Denotes minimum entropy at s_0.
$(\cdot)_o$	Denotes $\mathcal{H}_\infty$-optimal.
$(\cdot)_{\mathcal{L}_2}$	Denotes $\mathcal{L}_2$-optimal.
$(\cdot)_{LQG}$	Denotes LQG-optimal.
$\mathcal{F}(P, K)$	Lower linear fractional map of P and K (Equation (3.1)).
$C(H)$	LQG cost of the system H (Definition 2.4.1).
$C_T(H)$	Finite-time LQG cost of H (Remark 2.4.2).
$J(H;\gamma)$	Auxiliary LQG cost of H (Definition 5.2.1).
$I(H;\gamma;s_0)$	Entropy at s_0 of H (Definition 2.2.1).
$I(H;\gamma;\infty)$	Entropy at infinity of H (Definition 2.2.2).
$\Omega_T(\theta)$	Exponential-of-quadratic cost (Equation (6.2)).
$J_T(\gamma)$	Exponential-of-quadratic cost (Equation (6.3)).
$\mathbf{E}\{(\cdot)\}$	Expected value of $(\cdot)$.
$\mathrm{Var}\{(\cdot)\}$	Variance of $(\cdot)$.
$\max\{\cdot\}$	Largest element of the set $\{\cdot\}$.

C.2 Acronyms

CARE	Control Algebraic Riccati Equation (Equation (7.3)).
FARE	Filter Algebraic Riccati Equation (Equation (7.4)).
HCARE	$\mathcal{H}_\infty$ Control Algebraic Riccati Equation (Equation (7.9)).
HFARE	$\mathcal{H}_\infty$ Filter Algebraic Riccati Equation (Equation (7.10)).
LEQG	Linear Exponential-of-Quadratic Gaussian.
LHS	Left-hand side.
LQG	Linear Quadratic Gaussian
RHS	Right-hand side.

Bibliography

[1] B. D. O. Anderson and D. L. Mingori. Use of frequency dependence in Linear Quadratic control problems to frequency-shape robustness. *Journal of Guidance Control and Dynamics*, 8(3):397–401, 1985.

[2] B. D. O. Anderson and J. B. Moore. *Linear Optimal Control.* Prentice-Hall, 1971.

[3] D.Z. Arov and M. G. Krein. Problem of search of the minimum entropy in indeterminate extension problems. *Functional Analysis and its Applications*, 15:123–126, 1981.

[4] D.Z. Arov and M. G. Krein. On the evaluation of entropy functionals and their minima in generalized extension problems. *Acta Scienta Mathematica*, 45:33–50, 1983. (In Russian).

[5] J. A. Ball and N. Cohen. Sensitivity minimization in $\mathcal{H}_\infty$-norm. Parametrization of all suboptimal solutions. *International Journal of Control*, 46(3):785–816, 1987.

[6] C. S. Ballantine. Products of positive definite matrices. III. *Journal of Algebra*, 10:174–182, 1968.

[7] A. Bensoussan and J. W. van Schuppen. Optimal control of partially observable stochastic systems with an exponential-of-integral performance index. *SIAM Journal on Control and Optimization*, 23(4):599–613, 1985.

[8] D. S. Bernstein and W. M. Haddad. LQG control with an $\mathcal{H}_\infty$ performance bound: a Riccati equation approach. In *Proceedings of the American Control Conference*, Atlanta, GA, 1988.

[9] D. S. Bernstein and W. M. Haddad. LQG control with an $\mathcal{H}_\infty$ performance bound: a Riccati equation approach. *IEEE Transactions on Automatic Control*, 34(3):293–305, 1989.

[10] S. Boyd, V. Balakrishnan, and P. Kabamba. On computing the $\mathcal{H}_\infty$-norm of a transfer matrix. In *Proceedings of the American Control Conference*, Atlanta, GA, 1988.

[11] J. P. Burg. *Maximum Entropy Spectral Analysis.* PhD thesis, Stanford University, 1975.

[12] C. C. Chu, J. C. Doyle, and E. B. Lee. The general distance problem in $\mathcal{H}_\infty$ optimal control theory. *International Journal of Control*, 44(2):565–596, 1986.

[13] M. H. A. Davis. *Linear Estimation and Stochastic Control.* Chapman and Hall, 1977.

[14] D. F. Delchamps. A note on the analyticity of the Riccati metric. In C. I. Byrnes and C. F. Martin, editors, *Lecture Notes in Applied Mathematics*, pages 37–42. American Mathematical Society, 1980.

[15] J. C. Doyle. Guaranteed margins for LQG regulators. *IEEE Transactions on Automatic Control*, 23(4):756–757, 1978.

[16] J. C. Doyle and C. C. Chu. *Robust Control of Multivariable and Large Scale Systems.* Final Technical Report, Honeywell Systems and Research Center, 1986.

[17] J. C. Doyle, K. Glover, P. P. Khargonekar, and B. A. Francis. State-space solutions to standard $\mathcal{H}_2$ and $\mathcal{H}_\infty$ control problems. *IEEE Transactions on Automatic Control*, 34(8):831–847, 1989.

[18] J. C. Doyle and G. Stein. Multivariable feedback design: concepts for a classical/modern synthesis. *IEEE Transactions on Automatic Control*, 26(1):4–16, 1981.

[19] J. C. Doyle, J. E. Wall, and G. Stein. Performance and robustness analysis for structured uncertainty. In *Proceedings of the Conference on Decision and Control*, Orlando, FL, 1982.

[20] H. Dym. *J-Contractive Matrix Functions, Reproducing Kernel Hilbert Spaces and Interpolation*, volume 71 of *Regional Conference Series in Mathematics.* American Mathematical Society, 1989.

[21] H. Dym and I. Gohberg. A maximum entropy principle for contractive interpolants. *Journal of Functional Analysis*, 65:83–125, 1986.

[22] H. Dym and I. Gohberg. A new class of contractive interpolants and maximum entropy principles. In I. Gohberg, editor, *Topics in Operator Theory and Interpolation. Operator Theory: Advances and Applications*, volume 29, pages 117–150. Birkhäuser Verlag, 1988.

[23] D. F. Enns. *Model Reduction for Control System Design.* PhD thesis, Stanford University, 1984.

[24] B. A. Francis. *A Course in $\mathcal{H}_\infty$ Control Theory*, volume 88 of *Lecture Notes in Control and Information Sciences.* Springer–Verlag, 1987.

[25] F. R. Gantmacher. *The Theory of Matrices.* Chelsea Publishing Co., 1959.

[26] K. Glover. All optimal Hankel–norm approximations of linear multivariable systems and their $\mathcal{L}_\infty$–error bounds. *International Journal of Control*, 39(6):1115–1193, 1984.

[27] K. Glover. Minimum entropy and risk-sensitive control: the continuous time case. In *Proceedings of the Conference on Decision and Control*, Tampa, Fl, 1989.

[28] K. Glover and J. C. Doyle. State-space formulae for all stabilizing controllers that satisfy an $\mathcal{H}_\infty$-norm bound and relations to risk sensitivity. *Systems and Control Letters*, 11:167–172, 1988.

[29] K. Glover and J. C. Doyle. A state-space approach to $\mathcal{H}_\infty$ optimal control. In H. Nijmeijer and J. M. Schumacher, editors, *Three Decades of Mathematical System Theory: A Collection of Surveys at the Occasion of the 50th Birthday of Jan C. Willems*, volume 135 of *Lecture Notes in Control and Information Sciences.* Springer–Verlag, 1989.

[30] K. Glover, D. J. N. Limebeer, J. C. Doyle, E. M. Kasenally, and M. G. Safonov. A characterization of all solutions to the four block general distance problem. *To appear in SIAM Journal on Control and Optimization*, 1990?

[31] I. Gohberg, P. Lancaster, and L. Rodman. On Hermitian solutions of the symmetric algebraic Riccati equation. *SIAM Journal on Control and Optimization*, 24(6):1323–1334, 1986.

[32] R. A. Horn and C. R. Johnson. *Matrix Analysis.* Cambridge University Press, 1985.

[33] D. C. Hyland and D. S. Bernstein. The optimal projection equations for fixed-order dynamic compensation. *IEEE Transactions on Automatic Control*, 29(11):1034–1037, 1984.

[34] D. H. Jacobson. Optimal stochastic linear systems with exponential criteria and their relation to deterministic differential games. *IEEE Transactions on Automatic Control*, 18(2):124–131, 1973.

[35] E. A. Jonckheere and L. M. Silverman. A new set of invariants for linear systems—application to reduced order compensator design. *IEEE Transactions on Automatic Control*, 28(10):953–964, 1983.

[36] T. Kailath. *Linear Systems.* Prentice-Hall, 1980.

[37] V. Kučera. *Discrete Linear Control: The Polynomial Equation Approach.* Wiley, 1979.

[38] H. Kwakernaak and R. Sivan. *Linear Optimal Control Systems.* Wiley, 1972.

[39] D. J. N. Limebeer and Y. S. Hung. An analysis of the pole-zero cancellations in $\mathcal{H}_\infty$-optimal control problems of the first kind. *SIAM Journal on Control and Optimization*, 25(6):1457–1493, 1987.

[40] D. C. McFarlane and K. Glover. *Robust Controller Design Using Normalized Coprime Factor Plant Descriptions*, volume 138 of *Lecture Notes in Control and Information Sciences.* Springer–Verlag, 1990.

[41] D. G. Meyer. *Model Reduction via Fractional Representation.* PhD thesis, Stanford University, 1987.

[42] D. G. Meyer and G. F. Franklin. A connection between Linear Quadratic Regulator theory and normalized coprime factorizations. *IEEE Transactions on Automatic Control*, 32(3):227–228, 1987.

[43] D.G. Meyer. A fractional approach to model reduction. In *Proceedings of the American Control Conference*, Atlanta, GA, 1988.

[44] B. C. Moore. Principal component analysis in linear systems: controllability, observability and model reduction. *IEEE Transactions on Automatic Control*, 26(1):17–32, 1981.

[45] C. N. Nett, C. A. Jacobson, and M. J. Balas. A connection between state-space and doubly coprime fractional representations. *IEEE Transactions on Automatic Control*, 29(9):831–832, 1984.

[46] R. J. Ober and D. C. McFarlane. Balanced canonical forms for minimal systems: a normalized coprime factor approach. *Linear Algebra and Its Applications*, 122-124:23–64, 1989.

[47] Ph. Opdenacker and E. A. Jonckheere. LQG balancing and reduced LQG compensation of symmetric passive systems. *International Journal of Control*, 41(1):73–109, 1985.

[48] A. Ostrowski and H. Schneider. Some theorems on the inertia of general matrices. *Journal of Mathematical Analysis and Applications*, 4:72–84, 1962.

[49] L. Pernebo and L. M. Silverman. Model reduction via balanced state-space representations. *IEEE Transactions on Automatic Control*, 27(2):382–387, 1982.

[50] A. C. Ran and L. Rodman. On parameter dependence of solutions of algebraic Riccati equations. *Mathematics of Control, Signals, and Systems*, 1(3):269–284, 1988.

[51] R. M. Redheffer. On a certain linear fractional transformation. *Journal of Mathematics and Physics*, 39:269–286, 1960.

[52] W. Rudin. *Real and Complex Analysis.* Mc Graw-Hill, third edition, 1986.

[53] M. G. Safonov, A. J. Laub, and G. L. Hartman. Feedback properties of multivariable systems: the role and use of the return difference matrix. *IEEE Transactions on Automatic Control*, 26(1):47–65, 1981.

[54] C. E. Shannon and W. Weaver. *The Mathematical Theory of Communication.* University of Illinois Press, 1949.

[55] G. Stein and M. Athans. The LQG/LTR procedure for multivariable feedback control design. *IEEE Transactions on Automatic Control*, 32(2):105–113, 1987.

[56] M. Vidyasagar. *Control Systems Synthesis: A Factorization Approach.* MIT Press, 1985.

[57] P. Whittle. Risk-sensitive Linear/Quadratic/Gaussian control. *Advances in Applied Probability*, 13:764–777, 1981.

[58] P. Whittle. Entropy-minimizing and risk-sensitive control rules. *Systems and Control Letters*, 13:1–7, 1989.

[59] P. Whittle and J. Kuhn. A Hamiltonian formulation of risk-sensitive Linear/Quadratic/Gaussian control. *Internation Journal of Control*, 43:1–12, 1986.

[60] J. C. Willems. Least-squares stationary optimal control and the algebraic Riccati equation. *IEEE Transactions on Automatic Control*, 16(6):621–634, 1971.

[61] D. A. Wilson. Optimum solution of model-reduction problem. *Proceedings of the IEE*, 117(6):1161–1165, 1970.

[62] H. K. Wimmer. Monotonicity of maximal solutions of algebraic Riccati equations. *Systems and Control Letters*, 5:317–319, 1985.

[63] W. M. Wonham. *Linear Multivariable Control: A Geometric Approach.* Springer–Verlag, 1979.

[64] A. D. C. Youla, H. A. Jabr, and J. J. Bongiorno. Modern Wiener-Hopf design of optimal controllers. Part II—the multivariable case. *IEEE Transactions on Automatic Control*, 21(3):319–338, 1976.

[65] N. Young. *An Introduction to Hilbert Space.* Cambridge University Press, 1988.

[66] G. Zames. On the input-output stability of time-varying nonlinear feedback systems, Part I. *IEEE Transactions on Automatic Control*, 11(2):228–238, 1966.

[67] G. Zames. Feedback and optimal sensitivity: model reference transformations, multiplicative seminorms, and approximate inverses. *IEEE Transactions on Automatic Control*, 26(2):301–320, 1981.

Index

Lecture Notes in Control and Information Sciences

Edited by M. Thoma and A. Wyner

Vol. 97: I. Lasiecka/R. Triggiani (Eds.)
Control Problems for Systems
Described by Partial Differential Equations
and Applications
Proceedings of the IFIP-WG 7.2
Working Conference
Gainesville, Florida, February 3-6, 1986
VIII, 400 pages, 1987.

Vol. 98: A. Aloneftis
Stochastic Adaptive Control
Results and Simulation
XII, 120 pages, 1987.

Vol. 99: S. P. Bhattacharyya
Robust Stabilization Against
Structured Perturbations
IX, 172 pages, 1987.

Vol. 100: J. P. Zolésio (Editor)
Boundary Control and Boundary Variations
Proceedings of the IFIP WG 7.2 Conference
Nice, France, June 10-13, 1987
IV, 398 pages, 1988.

Vol. 101: P. E. Crouch,
A. J. van der Schaft
Variational and Hamiltonian
Control Systems
IV, 121 pages, 1987.

Vol. 102: F. Kappel, K. Kunisch,
W. Schappacher (Eds.)
Distributed Parameter Systems
Proceedings of the 3rd International Conference
Vorau, Styria, July 6–12, 1986
VII, 343 pages, 1987.

Vol. 103: P. Varaiya, A. B. Kurzhanski (Eds.)
Discrete Event Systems:
Models and Applications
IIASA Conference
Sopron, Hungary, August 3-7, 1987
IX, 282 pages, 1988.

Vol. 104: J. S. Freudenberg/D. P. Looze
Frequency Domain Properties of Scalar
and Multivariable Feedback Systems
VIII, 281 pages, 1988.

Vol. 105: Ch. I. Byrnes/A. Kurzhanski (Eds.)
Modelling and Adaptive Control
Proceedings of the IIASA Conference
Sopron, Hungary, July 1986
V, 379 pages, 1988.

Vol. 106: R. R. Mohler (Editor)
Nonlinear Time Series and
Signal Processing
V, 143 pages, 1988.

Vol. 107: Y. T. Tsay, L.-S. Shieh, St. Barnett
Structural Analysis and Design
of Multivariable Systems
An Algebraic Approach
VIII, 208 pages, 1988.

Vol. 108: K. J. Reinschke
Multivariable Control
A Graph-theoretic Approach
274 pages, 1988.

Vol. 109: M. Vukobratović/R. Stojić
Modern Aircraft Flight Control
VI, 288 pages, 1988.

Vol. 110: B. J. Daiuto, T. T. Hartley,
S. P. Chicatelli
The Hyperbolic Map and Applications
to the Linear Quadratic Regulator
VI, 114 pages, 1989

Vol. 111: A. Bensoussan, J. L. Lions (Eds.)
Analysis and Optimization
of Systems
XIV, 1175 pages, 1988.

Vol. 112: Vojislav Kecman
State-Space Models of Lumped
and Distributed Systems
IX, 280 pages, 1988

Vol. 113: M. Iri, K. Yajima (Eds.)
System Modelling and Optimization
Proceedings of the 13th IFIP Conference
Tokyo, Japan, Aug. 31 – Sept. 4, 1987
IX, 787 pages, 1988.

Vol. 114: A. Bermúdez (Editor)
Control of Partial Differential Equations
Proceedings of the IFIP WG 7.2
Working Conference
Santiago de Compostela, Spain, July 6–9, 1987
IX, 318 pages, 1989

Vol. 115: H.J. Zwart
Geometric Theory for Infinite
Dimensional Systems
VIII, 156 pages, 1989.

Vol. 116: M.D. Mesarovic, Y. Takahara
Abstract Systems Theory
VIII, 439 pages, 1989

Lecture Notes in Control and Information Sciences

Edited by M. Thoma and A. Wyner

Vol. 117: K.J. Hunt
Stochastic Optimal Control Theory with Application in Self-Tuning Control
X, 308 pages, 1989.

Vol. 118: L. Dai
Singular Control Systems
IX, 332 pages, 1989

Vol. 119: T. Başar, P. Bernhard
Differential Games and Applications
VII, 201 pages, 1989

Vol. 120: L. Trave, A. Titli, A. M. Tarras
Large Scale Systems:
Decentralization, Structure Constraints and Fixed Modes
XIV, 384 pages, 1989

Vol. 121: A. Blaquière (Editor)
Modeling and Control of Systems in Engineering, Quantum Mechanics, Economics and Biosciences
Proceedings of the Bellman Continuum Workshop 1988, June 13–14, Sophia Antipolis, France
XXVI, 519 pages, 1989

Vol. 122: J. Descusse, M. Fliess, A. Isidori, D. Leborgne (Eds.)
New Trends in Nonlinear Control Theory
Proceedings of an International Conference on Nonlinear Systems, Nantes, France, June 13–17, 1988
VIII, 528 pages, 1989

Vol. 123: C. W. de Silva, A. G. J. MacFarlane
Knowledge-Based Control with Application to Robots
X, 196 pages, 1989

Vol. 124: A. A. Bahnasawi, M. S. Mahmoud
Control of Partially-Known Dynamical Systems
XI, 228 pages, 1989

Vol. 125: J. Simon (Ed.)
Control of Boundaries and Stabilization
Proceedings of the IFIP WG 7.2 Conference Clermont Ferrand, France, June 20–23, 1988
IX, 266 pages, 1989

Vol. 126: N. Christopeit, K. Helmes M. Kohlmann (Eds.)
Stochastic Differential Systems
Proceedings of the 4th Bad Honnef Conference June 20–24, 1988
IX, 342 pages, 1989

Vol.127: C. Heij
Deterministic Identification of Dynamical Systems
VI, 292 pages, 1989

Vol. 128: G. Einarsson, T. Ericson, I. Ingemarsson, R. Johannesson, K. Zigangirov, C.-E. Sundberg
Topics in Coding Theory
VII, 176 pages, 1989

Vol. 129: W. A.Porter, S. C. Kak (Eds.)
Advances in Communications and Signal Processing
VI, 376 pages, 1989.

Vol. 130: W. A. Porter, S. C. Kak, J. L. Aravena (Eds.)
Advances in Computing and Control
VI, 367 pages, 1989

Vol. 131: S. M. Joshi
Control of Large Flexible Space Structures
IX, 196 pages, 1989.

Vol. 132: W.-Y. Ng
Interactive Multi-Objective Programming as a Framework for Computer-Aided Control System Design
XV, 182 pages, 1989.

Vol. 133: R. P. Leland
Stochastic Models for Laser Propagation in Atmospheric Turbulence
VII, 145 pages, 1989.

Vol. 134: X. J. Zhang
Auxiliary Signal Design in Fault Detection and Diagnosis
XII, 213 pages, 1989

Vol. 135: H. Nijmeijer, J. M. Schumacher (Eds.)
Three Decades of Mathematical System Theory
A Collection of Surveys at the Occasion of the 50th Birthday of Jan C. Willems
VI, 562 pages, 1989

Vol. 136: J. Zabczyk (Ed.)
Stochastic Systems and Optimization
Proceedings of the 6th IFIP WG 7.1 Working Conference, Warsaw, Poland, September 12–16, 1988
VI, 374 pages. 1989

Lecture Notes in Control and Information Sciences

Edited by M. Thoma and A. Wyner